The VEGAN PANTRY

성시우 (Sung Siwoo)

미쉐린 가이드 서울 | 부산 2025 1스타 비건 레스토랑 '레굠'의 총괄 셰프.
창의적인 식물성 요리를 연구하며 지속 가능한 미식을 탐구하고 있다.

The VEGAN PANTRY

일상 속 식물성 요리를 위한 첫걸음

초판 1쇄 발행 2025년 3월 20일
초판 2쇄 발행 2025년 4월 7일

지은이 성시우 | **영문 번역** 김유진 | **요리도움** 이해욱, 구모준, 이우주
펴낸이 박윤선 | **발행처** (주)더테이블

기획 박윤선 | **책임편집** 김영란 | **디자인** 김보라 | **사진** 신동민
영업·마케팅 김남권, 문성빈, 조용훈 | **경영지원** 김효선, 이정민

주소 경기도 부천시 조마루로385번길 122 삼보테크노타워 2002호
홈페이지 www.icoxpublish.com | **쇼핑몰** www.baek2.kr (백두도서쇼핑몰) | **인스타그램** @thetable_book
이메일 thetable_book@naver.com | **전화** 032) 674-5685 | **팩스** 032) 676-5685
등록 2022년 8월 4일 제 386-2022-000050 호 | **ISBN** 979-11-92855-17-2 (13590)

The VEGAN PANTRY

Your First Step
into Everyday Plant-Based Cooking

일상 속 식물성 요리를 위한 첫걸음

Sung Siwoo

성시우 지음

PROLOGUE

특별하지 않은 채식을 꿈꾸며

시대가 변하면서 식사가 단순한 한 끼 식사의 개념을 넘어서고 있습니다. 대중은 맛있는 음식 이상의 경험을 원하고 있고, 저 역시 셰프로서 새로운 영역을 개척해 나가고 싶은 마음이 있었습니다. 서로의 열망을 충족시킬 수 있는 요리가 무엇일지 고민하던 끝에 제가 내린 결론은 바로 채식이었습니다. 사실 채식이 친환경적이고 윤리적인 식단이라는 사실에는 이견이 없지만, 한국에서는 아직 채식에 대한 인식이 그다지 좋지 않은 것 같습니다. 영양이나 포만감이 부족해 건강하지 않을 것이라는 편견은 물론, 환경보호나 동물권을 위해 채식을 하는 사람을 유난스럽다고 여기는 사회적 시선도 여전합니다. 채식 인구가 증가하고 시장이 성장하고 있다고는 하지만, 한국에서 채식은 여전히 주류와는 거리가 먼, 유별난 음식으로 간주되고 있습니다. 이러한 고정 관념 때문에 채식을 실천하고자 했던 사람들도 얼마 가지 않아 포기하게 되는 것 같아 안타까운 마음이었습니다.

저는 셰프로서 채식의 아름다움을 더 많은 이들에게 알리기 위해 샐러드나 사찰음식, 대체육 패스트푸드에 한정된 국내 채식 시장에 조금 더 다양한 대안을 제시하고 싶었습니다. 자연 식물성 재료만으로도 훌륭한 미식 경험을 선사할 수 있다면 사람들의 인식을 바꿀 수 있을 거라고 확신했습니다. 채식이 특별한 이유가 있어야만 먹는 식단이 아니라, 평범한 하나의 일상식이 될 수 있기를 바라는 마음으로 채식 레스토랑을 구상하게 되었습니다. 그 시작은 2017년, 2018년 두 차례 선보인 팝업 레스토랑 '레귬스(LÉGUMES)'였습니다. '건강을 위한 혹은 신념에 의한 것이 아닌, 그 자체로 맛있고 자연스러운 채식'이라는 슬로건 아래, 채식을 다이닝 코스로도 운영해 보고 캐주얼한 한상차림 방식으로도 운영해 보면서 고객으로부터 다양한 의견을 수렴할 수 있었습니다. 이렇게 오랜 시행착오 끝에 완전 채식 다이닝의 가능성을 확인할 수 있었고, 2023년 마침내 100% 식물성 기반 레스토랑 '레귬(LÉGUME)'을 오픈하게 된 것입니다.

레귬은 단순히 채소 음식을 선보이는 레스토랑이 아니라, 채식을 통해 사람들과 소통하면서 더 나은 식사 경험을 만들어가는 공간입니다. 잘 알려지지 않은 품종을 발굴하거나 창의적인 조리법을 개발하여 채소가 가진 가능성을 새로이 발견하고, 이를 통해 남녀노소 모두가 즐길 수 있는 식사 경험을 제공하는 것이 목표입니다. 주위의 많은 우려에도 묵묵히 정진한 결과, '2025 서울 미식 100선(100 Taste of Seoul 2025)'

과 '미쉐린 가이드 서울 | 부산 2025(The MICHELIN Guide Seoul & Busan 2025)'에 이름을 올리는 등 국내외로 좋은 평가를 받으며 한국적인 채식을 세계에 알리는 데 힘쓰고 있습니다. 레귬은 채식에 친환경 실천뿐만 아니라 사회적 상생의 개념을 접목하여 채식에 담긴 지속 가능한 가치를 자연스럽게 환기하고 있습니다. 자투리 채소나 과일 껍질을 활용하는 레시피를 통해 제로 웨이스트(zero waste)를 추구하고, 친환경 작품을 만드는 작가들과의 협업으로 만든 업사이클링(upcycling) 식기를 사용하며, 지역 농가에서 직접 수확해 온 채소를 활용한 제철 음식을 만들어 우리 농산물의 우수성을 알리고 있습니다. 이는 고객이 레스토랑에서 식사하는 과정 속에서 자연스럽게 지속 가능한 생활 방식을 경험해 볼 수 있도록 설계한 것입니다. 레귬은 이처럼 건강한 삶의 방식을 제안하고 이를 사람들과 공유하면서 지속 가능한 문화를 만들어가고자 합니다.

이번 기회에 제가 추구하는 음식과 가치를 여러분과 공유할 수 있게 되어 진심으로 기쁩니다. 『더 비건 팬트리』는 셰프로서 그동안의 고민과 노력이 고스란히 담겨 있는 책입니다. 채식에 관한 요리 철학과 노하우를 레시피에 녹여내 채식의 교과서 같은 책을 만들고 싶었습니다. 집에서도 어렵지 않게 만들 수 있고 다 함께 즐길 수 있는 대중적인 음식으로 접근해 보았습니다. 채소 요리의 기본이 되는 베이직 레시피부터 이를 응용해서 만들 수 있는 요리의 예시까지 다채롭게 담아냈습니다. 책에 나온 요리를 하나씩 따라 해 보면서 무궁무진한 영감을 얻어 여러분만의 개성이 담긴 음식도 연구해 보기를 바랍니다. 그때 비로소 제가 의도했던 지속 가능성의 진정한 의미가 완성될 수 있을 것으로 기대합니다. 사실 제 이름을 건 요리책을 출간하는 일은 셰프로서 오랜 꿈이자 제 인생의 버킷리스트 중 하나였습니다. 그 꿈을 이룰 수 있게 해준 우리 레귬 팀 해욱, 모준, 우주에게 항상 고맙고, 이 책이 나오기까지 많은 도움 주신 더테이블 관계자분들께도 감사의 말씀을 전하고 싶습니다. 이 책으로 말미암아 한국에서도 채식이 특별하지 않은 일상식 중 하나로 자리 잡는 데 조금이나마 도움이 될 수 있으면 좋겠습니다. 여러분도 이 책을 통해 채식을 바라보는 시각에 약간의 변화가 생겼기를, 그리고 그 변화가 일상 속에서 작지만 의미있는 실천을 이끌어낼 수 있기를 바랍니다. 고맙습니다.

2025년 2월

성시우

PROLOGUE

Nothing Extraordinary: Dreaming of Plant-based Cuisine

As times continue to change, so too does the concept of a meal, going beyond mere nourishment. People today want more than just delicious food, and as a chef I, too, crave the opportunity to explore new horizons. After reflecting on what kind of cooking could fulfill these evolving aspirations, I found my answer in plant-based cuisine. While a vegetarian diet is undeniably ethical and eco-conscious, it still faces an unfavorable perception in Korea. Prejudice against vegetarian food frames it as unhealthy due to perceived deficiencies in nutrition or satiety, while societal biases see those who adopt it for animal rights or environmental reasons as overly particular. Despite a growing vegetarian population and market, plant-based food remains on the fringe and is often regarded as peculiar in Korea. I find it regrettable that such misconceptions deter people from sustaining a vegetarian diet.

As a chef, I wanted to showcase the true beauty of plant-based cuisine by broadening Korea's limited vegan dining scene, which had long been confined to salads, temple cuisine, or fast food with meat-substitutes. I was confident that public perception would change if I could offer an exceptional gastronomic experience using only natural plant-based ingredients. Hoping to make plant-based cuisine a normal part of daily meals rather than something reserved for special occasions, I set out to envision a plant-based restaurant. The journey began with **LÉGUMES**, a pop-up restaurant we hosted twice in 2017 and 2018. Under the slogan "Delicious, natural vegetarian cuisine on its own—neither for health nor driven by belief," we explored different approaches each year, presenting plant-based dishes either as a dining course or a casual set menu. Through extensive trial and error, we gathered diverse guest feedback and validated the potential of a vegan dining restaurant, which ultimately led to the opening of **LÉGUME**, a 100% plant-cuisine restaurant in 2023.

LÉGUME is more than just a restaurant offering vegetable dishes; it is dedicated to enriching the dining experience by connecting with people through plant-based cuisine. Our aim is to unlock the full potential of vegetables by exploring lesser-known varieties and developing innovative cooking techniques, offering a refined dining experience that appeals to people of all ages and backgrounds. Despite numerous concerns, LÉGUME has earned both local and international recognition, with its name featured in "100 Taste of Seoul 2025" and "the MICHELIN Guide Seoul & Busan 2025." We keep pushing boundaries, striving to bring Korean-inspired vegan cuisine to the world. Sustainability is core to our philosophy at Legume, with our plant-based cuisine seamlessly integrated with both eco-friendly and social co-existence. We pursue zero waste through inventive recipes that repurpose vegetable remnants or fruit peels, and we collaborate with local artists to craft upcycled tableware from eco-conscious materials. Additionally, by using fresh, locally harvested produce in our seasonal cuisine, we proudly showcase the excellence of Korean agriculture. These initiatives enable our guests to effortlessly engage with a sustainable lifestyle during their dining experience. In this way, LÉGUME promotes a culture of sustainability, while also encouraging a healthier lifestyle and sharing it with a wider community.

I am truly delighted to share the food and values I have pursued through this opportunity. *The Vegan Pantry* is a true reflection of my journey—my struggles and experiences as a chef. My goal was to create a foundational guide to plant-based cuisine, weaving my culinary philosophy and practical expertise into every recipe. To make plant-based cuisine more accessible at home and enjoyable for everyone, I focused on familiar, popular dishes. The book offers a carefully curated selection of plant-based recipes, ranging from basic recipes to creative applications that build upon them. I hope it becomes an endless source of inspiration as you follow each recipe, ultimately guiding you to create your own dishes, infused with your unique touch; that is when I believe the true meaning of sustainability I envisioned will finally be fulfilled. Publishing a cookbook under my name has long been a dream of mine as a chef, and a cherished goal on my lifelong bucket list. I am deeply grateful to our team members—Haewook, Mojoon, and Wooju—for helping me bring this dream to life. I also extend my gratitude to The Table staff for their invaluable support in making this book a reality. I hope that this book, even in a small way, contributes to making plant-based cuisine an everyday part of Korean meals rather than something seen as extraordinary. More importantly, I hope it sparks even the slightest shift in your perspective—one that may encourage subtle, if meaningful, actions in your daily life. Thank you.

February 2025

Author **Sung Siwoo**

CONTENTS

BASIC RECIPES

DISHES

DESSERTS

BASIC RECIPES

01

VEGETABLE STOCK

채수

채수는 채소 요리에 가장 기본이 되는 요소입니다. 물 대신 채수를 사용하면 요리에 다채로운 풍미를 더할 수 있습니다.

Vegetable stock is a fundamental element of plant-based cuisine, bringing depth and enhancing flavor when used instead of plain water.

900ml 분량
Yields 900 ml

냉장 3일, 냉동 1달
Lasts for 3 days when refrigerated, or 1 month when frozen

INGREDIENTS

채수

차가운 물	1L
건다시마	20g
당근	100g
양파	200g
셀러리	100g
통후추	3알
정향	1알
월계수 잎	1장

Vegetable Stock

1 L	cold water
20 g	dried kelp
100 g	carrot
200 g	onion
100 g	celery
3	black peppercorns
1	clove
1	bay leaf

1

2

3

4

HOW TO MAKE

채수

1. 냄비에 차가운 물과 건다시마를 넣고 강불에서 끓이다가, 물이 끓으면 약불로 줄여 3분간 우린 후 다시마를 건진다.

TIP. 건다시마는 흐르는 물에 깨끗이 세척 후 사용한다.

2. 당근, 양파, 셀러리를 아주 잘게 다진다.

TIP. 당근, 양파, 셀러리는 흐르는 물에 세척 후 사용한다.
셀러리는 섬유질을 제거한 후 사용한다.
빠른 작업을 위해 푸드프로세서를 이용해도 된다.

3. **1**에 **2**와 통후추, 정향, 월계수 잎을 넣고 약불에서 7~8분간 끓인다.

4. **3**을 체에 내려 채수와 잔여 채소를 분리한다.

TIP. 걸러낸 잔여 채소는 채소 커리(19p)와 채소과자(169p)를 만들 때 사용한다.
완성된 채수는 충분히 식힌 후 냉장고 또는 냉동고에 보관한다.

Vegetable Stock

1. In a pot, combine the dried kelp and cold water, then bring to a boil over high heat. Once it reaches to a boil, reduce the heat to low, let the kelp infuse for 3 minutes, then remove it.

TIP. Rinse the dried kelp thoroughly under running water before use.

2. Finely chop the carrot, onion, and celery.

TIP. Rinse the vegetables under running water before use.
Peel the celery to remove its fibers before chopping.
A food processor can be used for quicker preparation.

3. Combine (**1**) and (**2**) in a pot, then add the black peppercorns, cloves, and bay leaf. Simmer over low heat for 7 to 8 minutes.

4. Strain (**3**) through a sieve to separate the stock from the vegetable solids.

TIP. Reserve the vegetable solids for making vegetable curry (p.19) and vegetable chips (p.169).
Let the vegetable stock cool before storing it in the refrigerator or freezer.

02

VEGETABLE CURRY

채소 커리

채수를 거르고 남은 채소를 활용해 만들어서 맛도 영양도 풍부한 채소 커리입니다. 음식물 쓰레기를 줄이는 제로 웨이스트 레시피로 친환경적인 생활 습관을 실천해 보세요.

Made with vegetables strained from vegetable stock, this curry is rich in both flavor and nutrients. Embrace an eco-friendly lifestyle with this zero-waste recipe that reduces food waste.

2인 분량
Serves 2

냉장 3일
Lasts for 3 days when refrigerated

INGREDIENTS

채소 커리

잔여 채소(17p)	300g
채수(15p)	약간
다진 마늘	10g
레드 커리 페이스트 (수리타이)	30g
식용유	약간
소금	약간
설탕	약 30g
후추	약간

Vegetable Curry

300 g	cooked vegetable solids (p.17)
Q.S.	vegetable stock (p.15)
10 g	chopped garlic
30 g	red curry paste (SUREE Thai Red Curry Paste)
Q.S.	vegetable oil
Q.S.	salt
About 30 g sugar	
Q.S.	ground black pepper

HOW TO MAKE

채소 커리

1. 잔여 채소를 믹서에 넣고 곱게 간다.

TIP. **채소가 잘 갈리지 않으면 채수를 조금씩 추가해 가면서 갈아준다.**

2. 팬에 **1**을 넣고 약불에서 뭉근히 졸여 원하는 농도로 맞춘다.
3. 다른 팬에 식용유를 두른 후, 다진 마늘을 넣고 향이 날 때까지 볶는다.
4. **3**에 레드 커리 페이스트를 넣고 볶는다.
5. **4**에 **2**를 넣고 잘 섞는다.
6. 약불에서 원하는 농도까지 끓인 후 소금, 설탕, 후추로 간한다.

TIP. **설탕은 기호에 맞게 가감해 사용한다.**
완성된 레드 채소 커리는 충분히 식힌 후 냉장 또는 냉동 보관한다.

Vegetable Curry

1. Place the cooked vegetables solids in a blender and blend until smooth.

TIP. **If the mixture is too thick to blend smoothly, gradually add the vegetable stock while blending.**

2. Transfer (**1**) to a pan and simmer over low heat until it reaches the desired consistency.
3. In a separate pan with vegetable oil, cook the chopped garlic until aromatic.
4. Add the red curry paste into (**3**), then cook and stir to combine.
5. Add (**2**) to (**4**) and stir until fully incorporated.
6. Simmer over low heat until it reaches the desired consistency, then season with salt, sugar, and black pepper.

TIP. **Adjust the sugar to your taste.**
Let the red vegetable curry cool before storing it in the refrigerator or freezer.

NAMUL PESTO

나물 페스토

한국의 나물은 저마다 고유한 향을 지니고 있습니다. 제철 나물로 비건 페스토를 만들어 두면 오랫동안 저장해 두고 다양한 요리에 활용할 수 있습니다. 한 가지 나물만 사용해도 되지만, 여러 나물을 섞어 만들면 독특한 맛과 향을 조합할 수 있습니다. 레시피에서는 시중에서 쉽게 구할 수 있는 참나물을 사용했습니다.

Each kind of Korean namul has its own distinct aroma. When made into pesto, seasonal namul can be preserved longer and used in a variety of dishes. While this vegan pesto can be made with a single type, combining multiple varieties enhances its distinctive flavor and aroma. This recipe features chamnamul, a widely available option in Korea.

800g 분량
Yields 800 g

냉장 3일
Lasts for 3 days when refrigerated

INGREDIENTS

나물 페스토

참나물	150g
엑스트라 버진 올리브오일	270g
마늘	2쪽
소금	10g
후추	6g
설탕	10g
영양효모	40g
레몬즙	20g
캐슈넛	180g

* 나물은 사람이 먹을 수 있는 풀이나 나뭇잎, 혹은 이를 양념하여 무친 음식으로, 이 책에서는 조리나 양념을 하지 않은 신선한 나물을 의미한다. 나물이 없는 경우에는 시금치나 케일, 루콜라 등 다른 잎채소로 대체할 수 있다.

Namul Pesto

150 g	chamnamul
270 g	extra virgin olive oil
2	garlic cloves
10 g	salt
6 g	ground black pepper
10 g	sugar
40 g	nutritional yeast
20 g	lemon juice
180 g	cashew nuts

* Namul refers to edible greens, leaves, or seasoned vegetable dishes made from them. In this book, it specifically means fresh greens or leaves that are uncooked and unseasoned. If namul is unavailable, substitute it with leafy greens like spinach, kale, or arugula.

1

2

3

HOW TO MAKE

나물 페스토

1. 참나물은 깨끗이 씻어 물기를 제거한 후 듬성듬성 자른다.
2. 믹서에 캐슈넛을 제외한 모든 재료를 넣고 곱게 간다.
3. 2에 캐슈넛을 넣고 또 한번 곱게 갈아 마무리한다.

TIP. **완성된 페스토는 한 번에 먹기 좋게 소분해서 냉동 보관한다.**

Namul Pesto

1. Rinse the chamnamul thoroughly, drain, and chop coarsely.
2. In a blender, combine all the ingredients except the cashew nuts and blend until finely ground.
3. Add the cashew nuts to (**2**) and continue blending until smooth.

TIP. **Divide the pesto into single-serving portions and freeze for easy use.**

SOY MILK MAYONNAISE

두유 마요네즈

마요네즈는 서양 요리에서 기본이 되는 소스로, 일반적으로는 달걀노른자를 유화시켜 만들지만 두유를 활용하면 더욱 고소한 비건 마요네즈를 완성할 수 있습니다.

Mayonnaise, an essential sauce in Western cooking traditionally made by emulsifying egg yolks, can be crafted vegan with soy milk for a richer, nuttier flavor.

486g 분량
Yields 486 g

냉장 3일(진공 보관 시 냉장 1주)
Lasts for 3 days when refrigerated
(or 1 week when vacuum-sealed and refrigerated)

INGREDIENTS

두유 마요네즈

무가당 두유	200g
셰리 와인 비네거	30g
디종 머스타드	30g
소금	6g
메이플 시럽	20g
카놀라유	200g

Soy Milk Mayonnaise

200 g	unsweetened soy milk
30 g	sherry wine vinegar
30 g	Dijon mustard
6 g	salt
20 g	maple syrup
200 g	canola oil (rapeseed oil)

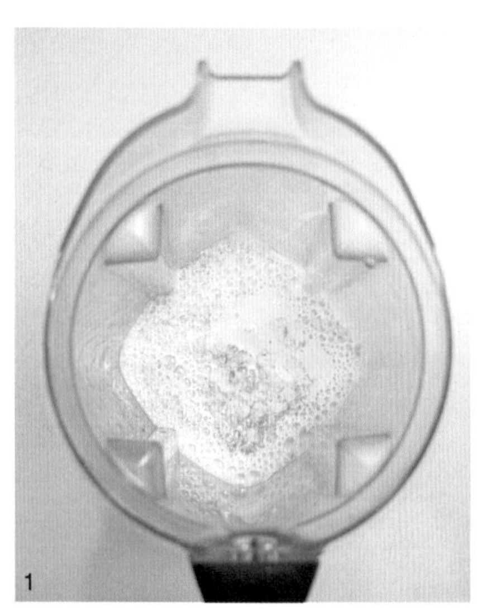
1

2

HOW TO MAKE

두유 마요네즈

1. 믹서에 카놀라유를 제외한 모든 재료를 넣고 간다.
2. 카놀라유를 세 번에 나누어 넣으면서 믹싱해 유화시킨다.

TIP. 분리된 기름이 보이지 않는 매끈한 질감이 되면 완성한다. 완성된 마요네즈는 밀폐 용기에 담거나 진공 포장하여 냉장 보관한다.

Soy Milk Mayonnaise

1. In a blender, combine all the ingredients except the canola oil and blend.
2. Continue blending while gradually adding the canola oil in three batches to emulsify.

TIP. Blend until smooth to ensure no visible oil separation. Store the mayonnaise in an airtight container or a vacuum-sealed bag, then refrigerate.

05

SOY MILK YOGURT

두유 요거트

두유를 유산균으로 발효해서 만든 비건 요거트입니다. 견과류나 과일을 곁들여 아침이나 디저트로 가볍게 즐겨보세요.

This vegan yogurt is made by fermenting soy milk with lactic acid bacteria. Enjoy it with nuts and fruits for a light breakfast or dessert.

700g 분량
Yields 700 g

냉장 3일
Lasts for 3 days when refrigerated

INGREDIENTS

두유 요거트

무가당 두유	950ml
비건 요거트 스타터	2g

Soy Milk Yogurt

950 ml	unsweetened soy milk
2 g	vegan yogurt starter

HOW TO MAKE

두유 요거트

1. 용기와 스푼을 깨끗하게 열탕 소독하여 준비한다.

TIP. **발효 시 잡균 발생 방지를 위해 소독이 가능한 유리 용기나 내열 실리콘 용기, 스푼을 사용한다.**

2. 소독한 용기에 상온의 무가당 두유를 담는다.
3. 비건 요거트 스타터를 넣고 소독한 스푼으로 잘 저어서 용해시킨다.
4. 랩으로 입구를 밀봉한 후 따듯한 곳(25°C~30°C)에서 약 12시간 발효시킨다.
5. 발효가 끝나면 면포를 얹은 체에 **4**를 붓고 그대로 냉장고에 넣어 요거트와 두유 유청을 분리한다.

TIP. **완성된 두유 요거트는 밀폐 용기에 담아 냉장 보관한다. 요거트가 너무 되직하면 분리한 두유 유청을 추가해 농도를 조절한다.**

Soy Milk Yogurt

1. Sterilize the container and spoon with boiling water, then set aside.

TIP. **Use a sterilizable glass or heat-resistant silicone container and a suitable spoon to prevent harmful germs from developing during fermentation.**

2. Pour the unsweetened soy milk at room temperature into the sterilized container.
3. Add the vegan yogurt starter and stir thoroughly with the sterilized spoon until fully dissolved.
4. Cover the container tightly with plastic wrap and let the mixture ferment in a warm place, 77-86°F (25-30°C), for about 12 hours.
5. Once fermented, pour (**4**) onto a piece of muslin (cheesecloth) set over a sieve to separate the yogurt from the soy milk whey.

TIP. **Store the soy milk yogurt in an airtight container and refrigerate.**
If the yogurt is too thick, adjust the consistency by adding the soy milk whey.

ALMOND RICOTTA CHEESE

아몬드 리코타치즈

아몬드를 곱게 갈아 한층 더 고소한 비건 리코타치즈입니다. 샐러드나 빵에 곁들여 견과류 풍미를 부드럽게 즐겨보세요.

Crafted from finely ground almonds, this vegan ricotta cheese offers a nuttier flavor. Enjoy its rich flavor and creamy texture on salads or spread on bread.

100g 분량
Yields 100 g

냉장 3일
Lasts for 3 days when refrigerated

INGREDIENTS

아몬드 리코타치즈

탈피 아몬드	100g
물	300g
소금	3g
레몬즙	60g

Almond Ricotta Cheese

100 g	peeled almonds
300 g	water
3 g	salt
60 g	lemon juice

HOW TO MAKE

아몬드 리코타치즈

1. 믹서에 탈피 아몬드와 물을 넣고 오랫동안 곱게 갈아 아몬드 밀크를 만든다.
2. 아몬드 밀크를 체에 걸러 냄비에 넣고 중불로 끓인다.
3. 아몬드 밀크가 끓기 시작하면 소금을 넣는다.
4. 덩어리가 지기 시작하면 레몬즙을 넣고 불에서 내린다.
5. 면포를 얹은 체에 **4**를 부어 치즈와 유청을 분리한다.
6. 분리된 치즈를 다시 한번 체에 내려 질감을 곱게 만든다.

TIP. **완성된 리코타치즈는 밀폐 용기에 담아 냉장고에 보관한다.**

Almond Ricotta Cheese

1. Combine the peeled almonds and water in a blender, then blend thoroughly for an extended time to make almond milk.
2. Strain the almond milk through a sieve into a pot, then bring to a boil over medium heat.
3. When the almond milk comes to a boil, add salt.
4. When the mixture begins to clump, add the lemon juice and remove from the heat.
5. Pour (**4**) onto a piece of muslin (cheesecloth) set over a sieve to separate the cheese from the whey.
6. Strain the cheese through the sieve again to achieve a smooth texture.

TIP. **Store the ricotta cheese in an airtight container and refrigerate.**

GREEN HERB OIL

그린 허브오일

녹색 허브의 향과 색을 그대로 담아내 음식에 포인트를 더해주는 허브오일입니다. 본 레시피에 사용한 딜뿐만 아니라 다양한 녹색 계열의 허브를 활용해 각기 다른 매력의 허브오일을 만들어 보세요.

Green herb oil preserves the fresh aroma and vibrant color of green herbs, adding a bright accent to any dish. Customize your own version by blending various green herbs in addition to the dill used in this recipe.

400g 분량
Yields 400 g

냉장 3일
Lasts for 3 days when refrigerated

INGREDIENTS

그린 허브오일

딜	200g
카놀라유	450g

* 다른 녹색 계열의 허브로 대체 가능하다.

Green Herb Oil

200 g	fresh dill
450 g	canola oil (rapeseed oil)

* Other green herbs can be used as substitutes.

1
2
3
4
5

HOW TO MAKE

그린 허브오일

1. 끓는 물에 딜을 넣고 색이 선명해질 때까지 가볍게 데친 후 얼음물에 빠르게 담가 식힌다.
2. 딜을 건진 후 손으로 꼭 짜서 물기를 뺀다.
3. **2**를 듬성듬성 자른다.
4. 믹서에 **3**과 카놀라유를 넣고 고속으로 약 8분간 곱게 간다.
5. 체에 면포를 얹고 **4**를 부어 잔여물을 걸러낸다.

TIP. 완성된 그린 허브오일은 충분히 식힌 후 밀폐 용기에 담거나 진공 포장하여 냉동 보관한다.

Green Herb Oil

1. Lightly blanch the dill in boiling water until vibrant, then immediately plunge it into ice water to cool.
2. Remove the dill from the ice water and squeeze out excess moisture by hand.
3. Coarsely chop (**2**).
4. Combine the canola oil and (**3**) in a blender and blend at high speed until smooth for about 8 minutes.
5. Pour (**4**) onto a piece of muslin (cheesecloth) set over a sieve to strain out the solids.

TIP. Cool the green herb oil before storing it in an airtight container or a vacuum-sealed bag, then freeze.

MUSHROOM DUXELLES

양송이 뒥셀

양송이버섯과 다양한 채소를 잘게 다진 후 수분을 날려 가며 볶아 만든 양송이 뒥셀입니다. 버섯의 맛이 진하게 응축되어 있어 감칠맛이 좋고 다양한 채소 요리에 활용할 수 있습니다.

This mushroom duxelles is crafted by cooking finely chopped button mushrooms and assorted vegetables, while allowing the moisture to evaporate. With its rich, concentrated mushroom flavor, it boasts a deep umami profile and serves as a versatile addition to a variety of vegetable dishes.

500g 분량
Yields 500 g

냉장 3일
Lasts for 3 days when refrigerated

INGREDIENTS

양송이 뒥셀

양송이버섯	300g
양파	200g
당근	100g
셀러리	100g
타임	2g
식용유	약간
채수(15p)	100g
소금	약간
후추	약간
두유 마요네즈(27p)	20g

Mushroom Duxelles

300 g	button mushrooms
200 g	onion
100 g	carrot
100 g	celery
2 g	fresh thyme
Q.S.	vegetable oil
100 g	vegetable stock (p.15)
Q.S.	salt
Q.S.	ground black pepper
20 g	soy milk mayonnaise (p.27)

1
2
3
4
5
6

HOW TO MAKE

양송이 뒥셀

1. 양송이버섯은 겉껍질을 제거한 다음 기둥 부분까지 곱게 다진다.
2. 양파, 당근, 셀러리도 양송이버섯과 마찬가지로 곱게 다진다.

TIP. **양파, 당근, 셀러리는 흐르는 물에 세척 후 사용한다. 셀러리는 섬유질을 제거한 후 사용한다.**

3. 타임은 줄기를 제거하고 잎 부분만 다진다.
4. 약불로 예열한 팬에 식용유와 **2**를 넣고 수분을 날리면서 볶는다.
5. 채소의 수분이 날아가고 양파가 투명해지면 **1**을 추가해서 볶는다.

TIP. **채소가 바닥에 눌어붙지 않게 중간중간 채수를 조금씩 넣어가면서 볶는다.**

6. 채수의 수분이 모두 날아가고 페이스트 형태가 되면 믹싱볼에 옮겨 담은 후 **3**과 소금, 후추, 두유 마요네즈를 넣고 고루 섞어 완성한다.

TIP. **완성된 양송이 뒥셀은 충분히 식힌 후 밀폐 용기에 담아 냉장 또는 냉동 보관한다.**

Mushroom Duxelles

1. Peel the button mushrooms and finely chop them, including the stems.
2. Finely chop the onion, carrot, and celery.

TIP. **Rinse the vegetables under running water before use. Peel the celery to remove its fibers before chopping.**

3. Strip the thyme leaves from the stems and chop them.
4. In a preheated pan over low heat, cook (**2**) with vegetable oil, while allowing the moisture to evaporate.
5. Once the moisture evaporates from the vegetables and the onion turns transparent, add (**1**) and continue cooking.

TIP. **Gradually add the vegetable stock while cooking to prevent the mixture from sticking to the pan.**

6. When the vegetable stock has fully evaporated and the mixture thickens into a paste, transfer it to a mixing bowl. Add the soy milk mayonnaise, (**3**), salt, and black pepper, then mix thoroughly.

TIP. **Let the mushroom duxelles cool before storing it in an airtight container, then refrigerate or freeze.**

09

DRIED TOMATO & OLIVE PRESERVES

말린 토마토 & 올리브 절임

말린 토마토와 올리브를 향신료와 올리브오일에 절여 만든 별미입니다. 장기 보관도 가능하니 냉장고에 넣어 두었다가 가벼운 술안주나 다른 음식에 곁들여 즐겨보세요.

This delicacy is made by preserving olives and dried tomatoes in olive oil and spices for long-term storage. Keep refrigerated and enjoy as a side dish or a flavorful accompaniment to your drink.

630g 분량
Yields 630 g

냉장 3일
Lasts for 3 days when refrigerated

INGREDIENTS

말린 토마토 & 올리브 절임

코리앤더 씨드	10g
펜넬 씨드	10g
샬롯	60g
마늘	2쪽
로즈메리	5g
대추방울토마토	10알
오히블랑카 블랙올리브	310g
셰리 와인 비네거	100g
메이플 시럽	20g
엑스트라 버진 올리브오일	160g

Dried Tomato & Olive Preserves

10 g	coriander seeds
10 g	fennel seeds
60 g	shallot
2	garlic cloves
5 g	fresh rosemary
10	grape tomatoes
310 g	Hojiblanca black olives
100 g	sherry wine vinegar
20 g	maple syrup
160 g	extra virgin olive oil

1
2
3
4

HOW TO MAKE

말린 토마토 & 올리브 절임

1. 기름 없는 팬에 코리앤더 씨드와 펜넬 씨드를 가볍게 볶은 후 다진다.
2. 샬롯, 마늘, 로즈메리는 최대한 곱게 다진다.

TIP. **로즈메리는 가을, 겨울철처럼 억센 상태의 줄기가 아니면 잎과 줄기를 모두 다져 사용한다.**

3. 대추방울토마토는 세척 후 반으로 잘라 식품 건조기 또는 100°C 이하의 오븐에서 약 3시간 동안 건조한다.

TIP. **토마토는 수분이 남아 단단하지 않고 손가락으로 눌렀을 때 구부러지는 반건조 상태로 건조한다.**

4. **1**과 **2**가 담긴 믹싱볼에 수분을 제거한 토마토, 오히블랑카 블랙올리브, 셰리 와인 비네거, 메이플 시럽, 올리브오일을 넣은 후 고루 섞는다.
5. 완성된 토마토 & 올리브 절임을 밀폐 용기에 담아 냉장 보관한다.

Dried Tomato & Olive Preserves

1. Lightly dry-toast the coriander and fennel seeds, then chop them.
2. Chop the shallot, garlic, and rosemary as finely as possible.

TIP. **Use the entire rosemary sprig, including the leaves and stem, unless the stem is too tough, as it often is in autumn and winter.**

3. Rinse and halve the grape tomatoes, then dry them in a dehydrator or an oven preheated to below 212°F (100°C) for about 3 hours.

TIP. **Half-dry the tomatoes so they retain some moisture, remaining soft and bendable when gently pressed with your fingers.**

4. In a mixing bowl containing (**1**) and (**2**), add the dried tomatoes, Hojiblanca black olives, sherry wine vinegar, maple syrup, and olive oil. Mix thoroughly.
5. Transfer the dried tomato & olive preserves in an airtight container and refrigerate.

DISHES

CUCUMBER & QUINOA SALAD

오이 & 퀴노아 샐러드

살짝 절인 오이에 슈퍼푸드인 퀴노아를 조합해 산뜻하게 입맛을 돋우는 샐러드입니다. 과즙이 풍부한 멜론과 싱그러운 허브 소스가 조화롭게 어우러져 완벽한 균형감을 보여줍니다.

This salad invigorates your palate with a delightful combination of semi-pickled cucumber and superfood quinoa. Succulent melon and a vibrant herb sauce complete the dish in perfect balance.

1그릇 분량
Makes 1 dish

INGREDIENTS

루콜라 & 고수 소스

루콜라	100g
고수	80g
마늘	1쪽
물	50g
엑스트라 버진 올리브오일	50g
소금	약간
설탕	약간
후추	약간

오이 & 퀴노아 샐러드

퀴노아	240g
채수(15p)	300g
소금	약간
샬롯	1/2개
두유 마요네즈 (27p)	약간
후추	약간
미니 오이	2개

곁들임

멜론	200g
소금	약간
설탕	약간
엑스트라 버진 올리브오일	약간
라임즙	30g

플레이팅

레드 소렐	5장
핑크페퍼	1g
라임 제스트	약간

Arugula & Cilantro Sauce

100 g	fresh arugula
80 g	fresh cilantro
1	garlic clove
50 g	water
50 g	extra virgin olive oil
Q.S.	salt
Q.S.	sugar
Q.S.	ground black pepper

Cucumber & Quinoa Salad

240 g	quinoa
300 g	vegetable stock (p.15)
Q.S.	salt
1/2	shallot
Q.S.	soy milk mayonnaise (p.27)
Q.S.	ground black pepper
2	mini cucumbers

Garnishes

200 g	melon
Q.S.	salt
Q.S.	sugar
Q.S.	extra virgin olive oil
30 g	lime juice

Plating

5	fresh red sorrel leaves
1 g	crushed pink pepper
Q.S.	lime zest

Arugula & Cilantro Sauce

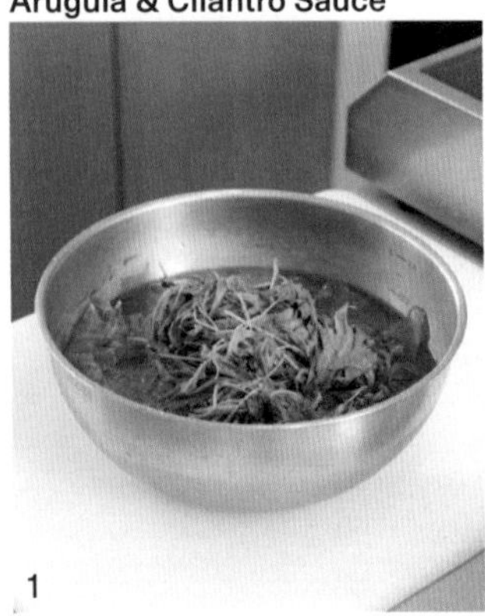
1

2

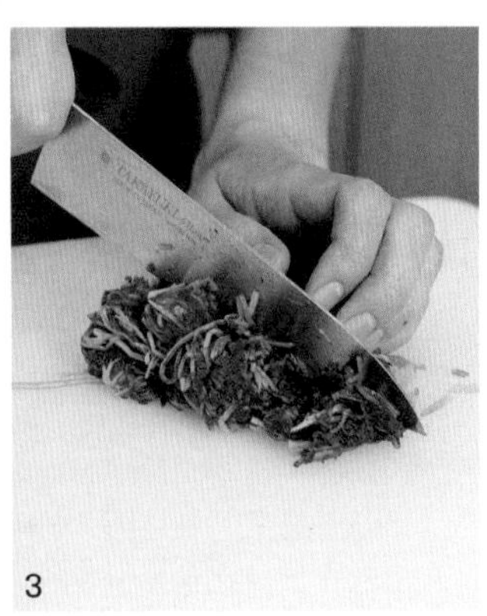
3

4

Cucumber & Quinoa Salad

1

2

3

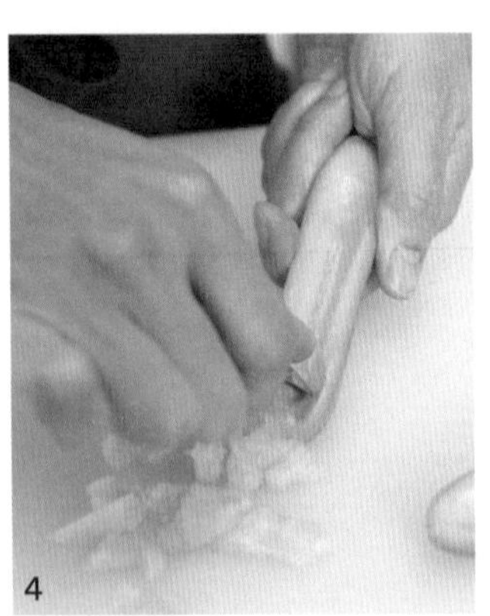
4

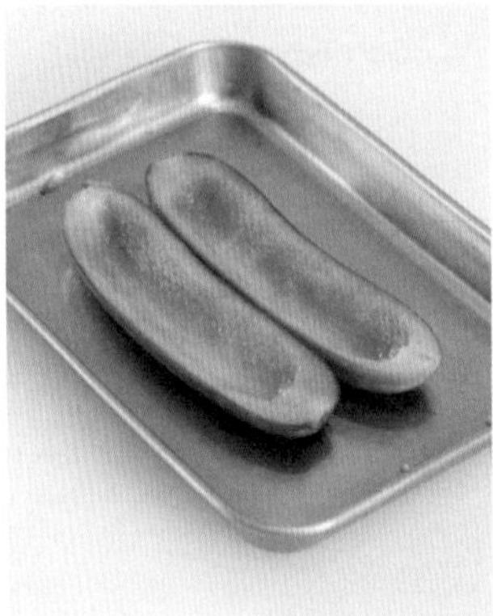

Garnishes

HOW TO MAKE

루콜라 & 고수 소스

1. 루콜라와 고수를 끓는 물에 가볍게 데친 후 얼음물에 빠르게 담가 식힌다.
2. 1을 건진 후 손으로 꼭 짜서 물기를 뺀다.
3. 2를 듬성듬성 자른다.
4. 믹서에 3과 나머지 재료를 모두 넣고 곱게 갈아 소스를 만든다.

Arugula & Cilantro Sauce

1. Lightly blanch the arugula and cilantro in boiling water, then immediately plunge them into ice water to cool.
2. Remove (**1**) from the ice water and squeeze out excess moisture by hand.
3. Coarsely chop (**2**).
4. Combine (**3**) and all the other ingredients in a blender, then blend until smooth to make the sauce.

오이 & 퀴노아 샐러드

1. 냄비에 퀴노아와 채수, 소금을 넣고 약불로 끓인 후 믹싱볼에 옮겨 담아 식힌다.

TIP. 퀴노아는 물에 씻은 후 부드러워질 때까지 익힌다.

2. 샬롯을 곱게 다진다.
3. 퀴노아가 담긴 믹싱볼에 2를 넣고, 두유 마요네즈, 소금, 후추를 더해 고루 버무린다.
4. 미니 오이는 반으로 잘라 씨를 파내고 소금을 고루 뿌려 10분간 절인 후 물로 씻어 낸다.

TIP. 물에 씻은 오이는 먹기 좋은 크기로 자른다.

Cucumber & Quinoa Salad

1. In a pot, simmer the quinoa with vegetable stock and salt over low heat, then transfer to a mixing bowl to cool.

TIP. Rinse the quinoa before cooking, then simmer until tender.

2. Finely chop the shallot.
3. Transfer (**2**) to the mixing bowl containing the cooled quinoa. Add the soy milk mayonnaise, salt, and black pepper, then toss thoroughly to combine.
4. Halve the mini cucumbers and scoop out the seeds. Sprinkle evenly with salt, let sit for 10 minutes, then rinse with water.

TIP. Cut the rinsed cucumbers into bite-sized pieces for easier eating.

곁들임

멜론을 먹기 좋은 크기로 잘라 소금, 설탕, 올리브오일, 라임즙으로 버무린다.

Garnishes

Cut the melon into bite-sized pieces for easier eating, then gently dress with salt, sugar, olive oil, and lime juice.

PLATING

1. 그릇에 오이 & 퀴노아 샐러드와 멜론을 조화롭게 담는다.
2. 1 위에 루콜라 & 고수 소스를 끼얹는다.
3. 레드 소렐과 핑크페퍼, 라임 제스트를 올려 마무리한다.

TIP. **레드 소렐은 찬물에 담가 싱싱하게 만들어 사용한다.**

1. Arrange the cucumber & quinoa salad with the dressed melon in a balanced presentation.
2. Spoon the arugula & cilantro sauce over (**1**).
3. Finish with the red sorrel, crushed pink pepper, and lime zest.

TIP. **Soak the red sorrel in ice water to refresh it before use.**

1

2

3

02

ALMOND RICOTTA CHEESE WITH APPLE & CELERY

아몬드 리코타치즈와 사과 & 셀러리

궁합이 좋은 사과와 셀러리로 만든 상큼한 샐러드입니다. 앞서 만든 아몬드 리코타치즈와 두유 마요네즈를 조합한 고소한 풍미의 소스와 함께 즐겨보세요.

This refreshing salad showcases the perfect pairing of apple and celery. Enjoy it with a savory sauce made from previously prepped almond ricotta cheese and soy milk mayonnaise.

1그릇 분량
Makes 1 dish

INGREDIENTS

아몬드 리코타치즈와 사과 & 셀러리

셀러리	1줄
사과	1/2개
엑스트라 버진 올리브오일	약간
소금	약간
후추	약간
아몬드 리코타치즈 (35p)	100g
두유 마요네즈(27p)	30g

플레이팅

구운 캐슈넛	6알
레드 소렐	약간
펜넬 잎	약간
그린 허브오일(39p)	5g

Almond Ricotta Cheese with Apple & Celery

1	celery rib
1/2	apple
Q.S.	extra virgin olive oil
Q.S.	salt
Q.S.	ground black pepper
100 g	almond ricotta cheese (p.35)
30 g	soy milk mayonnaise (p.27)

Plating

6	roasted cashew nuts
Q.S.	fresh red sorrel
Q.S.	fresh fennel leaves
5 g	green herb oil (p.39)

1
2
3
4
5

HOW TO MAKE

아몬드 리코타치즈와 사과 & 셀러리

1. 셀러리는 한입 크기로 썬다.

TIP. 셀러리는 흐르는 물에 세척 후 섬유질을 제거해 사용한다.

2. 사과를 한입 크기로 썬다.

TIP. 사과는 흐르는 물에 세척 후 사용한다.

3. 소금물에 **1**과 **2**를 20초 이내로 데친 후 얼음물에 빠르게 담가 식힌다.

TIP. 소금물은 물 1L, 소금 10g 비율로 끓여 사용한다.

4. **3**을 건져 키친타월로 물기를 제거한 후, 믹싱볼에 옮겨 올리브오일과 소금, 후추를 넣고 버무린다.

5. 다른 믹싱볼에 아몬드 리코타치즈와 두유 마요네즈를 섞어 소스를 만든다.

Almond Ricotta Cheese with Apple & Celery

1. Cut the celery into bite-sized pieces.

TIP. Rinse the celery under running water and peel to remove its fibers before cutting.

2. Cut the apple into bite-sized pieces.

TIP. Rinse the apple under running water before use.

3. Blanch (**1**) and (**2**) in boiling salted water for up to 20 seconds, then transfer immediately to ice water to cool.

TIP. For the salted water, use a ratio of 10 g of salt per 1 L of water and bring to a boil.

4. Drain (**3**) and pat dry with a kitchen towel. Transfer to a mixing bowl, then gently dress with olive oil, salt, and black pepper.

5. In a separate mixing bowl, mix the almond ricotta cheese and soy milk mayonnaise to make the sauce.

PLATING

1. 그릇에 아몬드 리코타치즈와 두유 마요네즈를 섞은 소스를 담은 후, 사과와 셀러리로 둘러싼다.
2. 사과와 셀러리 위에 구운 캐슈넛과 레드 소렐, 펜넬 잎을 올려 장식한다.

TIP. **레드 소렐과 펜넬 잎은 찬물에 담가 싱싱하게 만들어 사용한다.**

3. 숟가락 뒷부분으로 소스 중앙에 홈을 둥글게 판 후, 홈 안에 그린 허브오일을 담아 마무리한다.

1. Spoon the almond ricotta cheese and soy milk mayonnaise sauce onto a plate. Arrange the dressed apple and celery around the sauce.
2. Garnish with the roasted cashew nuts, red sorrel, and fennel leaves.

TIP. **Soak the red sorrel and fennel leaves in ice water to refresh them before use.**

3. Use the back of a spoon to create a round indentation in the center of the sauce, then fill it with the green herb oil.

1

2

3

03

KIWI SALAD WITH DILL DRESSING

키위 샐러드와 딜 드레싱

앞서 만든 그린 허브오일에 약간의 재료를 더하면 산뜻한 드레싱으로 변신합니다. 과일 샐러드에 허브 드레싱을 곁들여 재료 본연의 맛을 더욱 싱그럽게 즐겨보세요.

With just a few simple additions, previously prepped green herb oil transforms into a vibrant dressing. Serve it with a fruit salad to enhance and refresh its natural flavor.

1그릇 분량
Makes 1 dish

INGREDIENTS

딜 드레싱

양파	25g
딜오일(39p)	30g
라임즙	38g
소금	약간
후추	약간
설탕	약간

* 딜로 그린 허브오일(39p)을 만들어 사용한다.

Dill Dressing

25 g	onion
30 g	dill oil (p.39)
38 g	lime juice
Q.S.	salt
Q.S.	ground black pepper
Q.S.	sugar

* Make the green herb oil (p.39) with dill.

키위 샐러드

키위	1/2개
브로콜리니	3개

Kiwi Salad

1/2	kiwi
3	broccolini spears

플레이팅

아몬드 리코타치즈 (35p)	30g
베이비 루콜라	5g
딜	약간
레드 소렐	약간

Plating

30 g	almond ricotta cheese (p.35)
5 g	fresh baby arugula
Q.S.	fresh dill
Q.S.	fresh red sorrel

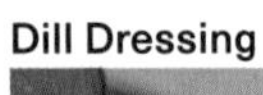

Dill Dressing

Kiwi Salad

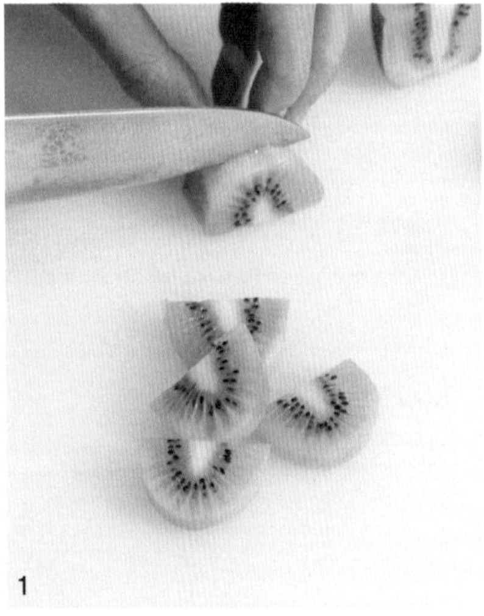

HOW TO MAKE

딜 드레싱

믹싱볼에 곱게 간 양파와 딜오일, 라임즙, 소금, 후추, 설탕을 넣고 섞는다.

Dill Dressing

In a mixing bowl, combine finely grated onion, dill oil, lime juice, salt, black pepper, and sugar, then stir until incorporated.

키위 샐러드

1. 키위는 껍질을 제거하여 먹기 좋은 크기로 자른다.
2. 브로콜리니는 줄기 부분의 껍질을 제거하고 소금물에 한번 데친 후 얼음물에 담가 아삭한 식감을 살린다.

TIP. **소금물은 물 1L, 소금 10g 비율로 끓여 사용한다.**

Kiwi Salad

1. Peel the kiwi and cut it into bite-sized pieces for easier eating.
2. Peel the broccolini stem, blanch the whole broccolini in salted boiling water, then transfer to ice water to enhance its crunchy texture.

TIP. **For the salted water, use a ratio of 10 g of salt per 1 L of water and bring to a boil.**

PLATING

1. 아몬드 리코타치즈를 커넬 모양으로 만들어 그릇에 담는다.
2. 아몬드 리코타치즈 우측에 키위와 브로콜리니, 베이비 루콜라, 딜, 레드 소렐을 담는다.

TIP. **베이비 루콜라, 딜, 레드 소렐은 찬물에 담가 싱싱하게 만든다.**

3. 아몬드 리코타치즈 좌측에 딜 드레싱을 곁들여 마무리한다.

1. Shape a quenelle of the almond ricotta cheese and place it on a plate.
2. Arrange the kiwi, broccolini, baby arugula, dill, and red sorrel to the right of the cheese.

TIP. **Soak the baby arugula, dill, and red sorrel in ice water to refresh them before use.**

3. Serve with the dill dressing placed to the left of the cheese.

1

2

3

04

ONION & BURDOCK ROOT SOUP

양파 & 우엉 수프

프렌치 어니언 수프(French onion soup)와 같은 방식으로 만들지만, 한식에서 많이 사용하는 우엉을 더해 자연스러운 단맛과 깊은 감칠맛을 한층 더 끌어올린 수프입니다.

Cooked using the same method as traditional French onion soup, this version brings out the onion's natural sweetness and deep umami, further enriched by the addition of burdock root, an ingredient commonly used in Korean cuisine.

1그릇 분량
Makes 1 dish

INGREDIENTS

양파 & 우엉 수프

양파	300g
우엉	100g
식용유	약간
채수(15p)	500g
소금	약간
후추	약간

Onion & Burdock Root Soup

300 g	onion
100 g	burdock root
Q.S.	vegetable oil
500 g	vegetable stock (p.15)
Q.S.	salt
Q.S.	ground black pepper

플레이팅

이탈리안 파슬리	약간

Plating

Q.S.	fresh Italian parsley

1
2
3
4
5
6
7

HOW TO MAKE

양파 & 우엉 수프

1. 양파를 최대한 얇게 채 썬다.
2. 우엉은 양파와 비슷한 크기로 채 썬다.

TIP. **우엉은 껍질째 깨끗이 씻어 사용한다.**

3. 식용유를 두른 냄비에 **1**을 넣고 약불로 볶아 캐러멜화시킨다.

TIP. **양파가 약 10분의 1 정도로 졸아들고 진한 밤색이 될 때까지 볶는다.**

4. 다른 냄비에 식용유를 두른 후 **2**를 넣고 볶는다.
5. 우엉이 익어서 향이 우러나면 **4**에 **3**과 채수를 넣는다.
6. 양파가 뭉치지 않게 살살 저어주며 약불에서 뭉근히 끓인다.
7. 재료의 맛이 충분히 우러나고 수프가 어느 정도 졸아들면 소금과 후추로 간해서 마무리한다.

Onion & Burdock Root Soup

1. Slice the onion as thinly as possible.
2. Slice the burdock root to a size similar to the onion slices.

TIP. **Rinse the burdock root thoroughly with the skin on before use.**

3. In a pot with vegetable oil, cook (**1**) over low heat until caramelized.

TIP. **Cook until the sliced onion reduces to one-tenth of its original volume and turns dark brown.**

4. In a separate pot, cook (**2**) in vegetable oil.
5. Once the sliced burdock root is cooked and aromatic, add (**3**) and vegetable stock to (**4**).
6. Simmer over low heat while gently stirring the caramelized onion to prevent clumping.
7. Once the soup develops a rich flavor from the ingredients and reduces to the desired consistency, season with salt and black pepper.

PLATING

1. 그릇에 따듯한 수프를 담는다.
2. 이탈리안 파슬리를 다져 올려 마무리한다.

1. Ladle the warm soup into a bowl.
2. Finish with chopped Italian parsley.

1

2

05

TOMATO SOUP

토마토 수프

'토마토 수프'라고 하면 대부분 빨간색 수프를 떠올리는데요. 이번에 소개할 수프는 토마토를 으깨지 않고 맑은 채수에 그 맛을 자연스럽게 우려내 익숙하면서도 색다른 느낌을 선사합니다.

"Tomato soup" often evokes the image of a red-colored dish. However, this version offers a familiar yet distinctive twist by steeping tomatoes in a clear broth rather than crushing them.

1그릇 분량
Makes 1 dish

INGREDIENTS

토마토 수프

방울토마토	10알
마늘	2쪽
양파	50g
템페	50g
식용유	약간
소금	약간
엑스트라 버진 올리브오일	약간
채수(15p)	280g

* 템페는 콩을 발효시켜 만든 인도네시아 전통 음식으로, 비건 요리에서 많이 활용된다.

Tomato Soup

10	cherry tomatoes
2	garlic cloves
50 g	onion
50 g	tempeh
Q.S.	vegetable oil
Q.S.	salt
Q.S.	extra virgin olive oil
280 g	vegetable stock (p.15)

* Tempeh, a traditional Indonesian food made from fermented soybeans, is widely used in vegan cuisine.

플레이팅

처빌	20g
처빌오일	약간

* 처빌로 그린 허브오일(39p)을 만들어 사용한다.

Plating

20 g	fresh chervil
Q.S.	chervil oil

* Make the green herb oil (p.39) with chervil.

HOW TO MAKE

토마토 수프

1. 방울토마토는 끓는 물에 10~20초간 데친 후 곧바로 얼음물에 담가 껍질을 제거하고 반으로 잘라 준비한다.
2. 마늘을 곱게 다진다.
3. 양파는 작은 주사위 모양으로 썬다.
4. 템페를 주사위 모양으로 썰어 식용유를 넉넉히 두른 냄비에 넣고 튀긴다.
5. 골고루 튀긴 템페는 소금으로 간한다.
6. 약불에 달군 냄비에 올리브오일을 두른 후 **2**을 넣고 볶는다.
7. 마늘 향이 우러나면 **3**을 넣고 볶는다.
8. 양파 향이 우러나면 **1**을 넣고 으깨지지 않도록 살살 볶는다.
9. **8**에 채수를 넣고 토마토가 무를 때까지 약불에서 뭉근히 끓인다.
10. 재료의 맛이 충분히 우러나고 채수가 어느 정도 졸아들면 불에서 내린다.

Tomato Soup

1. Blanch the cherry tomatoes in boiling water for 10 to 20 seconds, then immediately plunge them into ice water. Peel and halve them.
2. Finely chop the garlic.
3. Dice the onion into small pieces.
4. Dice the tempeh into pieces and fry in a pot with ample vegetable oil.
5. Season the evenly fried tempeh with salt.
6. In a preheated pot over low heat, cook (**2**) in olive oil.
7. Once the garlic becomes aromatic, add (**3**).
8. Once the onion becomes aromatic, add the tomatoes and stir gently to avoid crushing them.
9. Pour the vegetable stock into (**8**) and simmer over low heat until the tomatoes soften.
10. Once the stock develops a rich flavor from the ingredients and reduces to the desired consistency, remove from the heat.

PLATING

1. 그릇에 템페를 담는다.
2. 템페 위로 토마토 수프를 담는다.
3. 수프 위에 처빌을 다져 올리고 처빌오일을 뿌려 마무리한다.

1. Place the fried tempeh in a bowl.
2. Ladle the warm tomato soup over the tempeh.
3. Finish with a sprinkle of chopped chervil and a drizzle of chervil oil.

1

2

3

RED POTATO & ROSEMARY SOUP

홍감자 & 로즈메리 수프

속을 든든하게 채워주는 홍감자 수프입니다. 자칫 너무 담백하게 느껴질 수 있어 대파와 로즈메리, 올리브오일로 은은한 향을 더해 포인트를 주었습니다.

This satisfying red potato soup is enriched with scallion, rosemary, and olive oil, adding delicate aromas to elevate its flavor and prevent it from tasting too bland.

1그릇 분량
Makes 1 dish

INGREDIENTS

홍감자 & 로즈메리 수프

홍감자	200g
채수(15p)	300g
대파 흰부분	1줄
엑스트라 버진 올리브오일	약간
소금	약간
후추	약간

* 홍감자가 없으면 다른 분질감자류로 대체 가능하다.

Red Potato & Rosemary Soup

200 g	red potatoes
300 g	vegetable stock (p.15)
1	scallion (white part)
Q.S.	extra virgin olive oil
Q.S.	salt
Q.S.	ground black pepper

* If red potatoes are unavailable, other starchy potatoes can be used as substitutes.

플레이팅

로즈메리	5g
엑스트라 버진 올리브오일	약간

Plating

5 g	fresh rosemary
Q.S.	extra virgin olive oil

HOW TO MAKE

홍감자 & 로즈메리 수프

1. 홍감자는 소금물에 껍질째 삶아서 완전히 익힌다.

TIP. 소금물은 물 1L, 소금 10g 비율로 끓여 사용한다.
젓가락으로 찔렀을 때 부드럽게 들어갈 정도로 푹 익힌다.

2. 부드럽게 익힌 감자는 껍질을 제거한 후 뜨거울 때 체에 곱게 내린다.

3. 믹서에 **2**와 뜨겁게 끓인 채수를 넣고 약 10초간 간다.

TIP. 뜨거울 때 고속으로 갈면 위험하니 저속부터 시작하여 천천히 고속으로 갈아준다.

4. 냄비에 **3**을 다시 한번 체로 곱게 내린 후 소금, 후추를 넣고 가열한다.

5. 대파 흰 부분을 링 모양으로 채 썬다.

6. 올리브오일을 두른 팬에 **5**를 약불로 볶고 소금, 후추로 간한다.

Red Potato & Rosemary Soup

1. Fully cook the red potatoes, with the skin on, in boiling salted water.

TIP. For the salted water, use a ratio of 10 g of salt per 1 L of water and bring to a boil.
Cook until a chopstick easily pierces through.

2. Peel the cooked potatoes and press them through a fine-mesh sieve while hot.

3. Combine (**2**) and hot vegetable stock in a blender, then blend for about 10 seconds.

TIP. Since blending hot ingredients at high speed can be dangerous, start at low speed and gradually increase the speed to high.

4. Strain (**3**) through the sieve again into a pot, season with salt and black pepper, then bring to a boil.

5. Slice the white part of the scallion into rings.

6. Cook (**5**) in a pan with olive oil over low heat, then season with salt and black pepper.

PLATING

1. 그릇에 볶은 대파를 넣는다.
2. 따듯한 감자 수프를 담는다.
3. 수프 위에 로즈메리를 다져 올리고 올리브오일을 뿌려 마무리한다.

TIP. 가을, 겨울철처럼 억센 상태의 줄기가 아니면 잎과 줄기를 모두 다져 사용한다.

1. Place the cooked scallions in a bowl.
2. Ladle the warm potato soup over the scallions.
3. Finish with chopped rosemary and a drizzle of olive oil.

TIP. Use the entire rosemary sprig, including the leaves and stem, unless the stem is too tough, as it often is in autumn and winter.

1

2

3

07

TOASTED SOURDOUGH WITH DRIED TOMATO & OLIVE PRESERVES

사워도우 토스트와 말린 토마토 & 올리브 절임

바삭하게 구운 사워도우에 앞서 만든 토마토와 올리브 절임을 토핑으로 올려 오픈 샌드위치처럼 간편하게 즐겨보세요.

Top crispy-toasted sourdough with previously prepped tomato & olive preserves for an easy open-faced sandwich.

1그릇 분량
Makes 1 dish

INGREDIENTS

사워도우 토스트

청포도	3알
사워도우	100g
식용유	약간
마늘	1쪽
로즈메리	1줄
소금	약간

Toasted Sourdough

3	green grapes
100 g	sourdough
Q.S.	vegetable oil
1	garlic clove
1	fresh rosemary sprig
Q.S.	salt

플레이팅

두유 마요네즈(27p)	30g
말린 토마토 & 올리브 절임(47p)	50g
아몬드 리코타치즈(35p)	30g
딜	10g
처빌	10g

Plating

30 g	soy milk mayonnaise (p.27)
50 g	dried tomato & olive preserves (p.47)
30 g	almond ricotta cheese (p.35)
10 g	fresh dill
10 g	fresh chervil

1
2
3
4

HOW TO MAKE

사워도우 토스트

1. 청포도를 반으로 자른다.
2. 사워도우를 먹기 좋은 크기로 자른 후, 팬에 식용유를 넉넉히 두르고 앞뒤로 굽는다.
3. 사워도우에 노릇한 색이 돌면 손으로 으깬 마늘과 로즈메리를 넣어 향을 입힌다.
4. 소금으로 간해 마무리한다.

Toasted Sourdough

1. Halve the green grapes.
2. Cut the sourdough into bite-sized pieces for easier eating and toast in a pan with ample vegetable oil, flipping occasionally.
3. Once golden brown, add crushed garlic and rosemary to infuse the bread with their flavors.
4. Season with a pinch of salt.

PLATING

1. 사워도우 단면에 두유 마요네즈를 발라 그릇에 올린다.
2. 말린 토마토 & 올리브 절임을 올린다.
3. 반으로 자른 청포도를 올린다.
4. 아몬드 리코타치즈와 딜, 처빌을 올려 마무리한다.

TIP. **딜과 처빌은 찬물에 담가 싱싱하게 만들어 사용한다.**

1. Spread the soy milk mayonnaise on the toasted sourdough and place it on a plate.
2. Top with the dried tomato & olive preserves.
3. Garnish with the halved green grapes.
4. Finish with the almond ricotta cheese, dill, and chervil.

TIP. **Soak the dill and chervil in ice water to refresh them before use.**

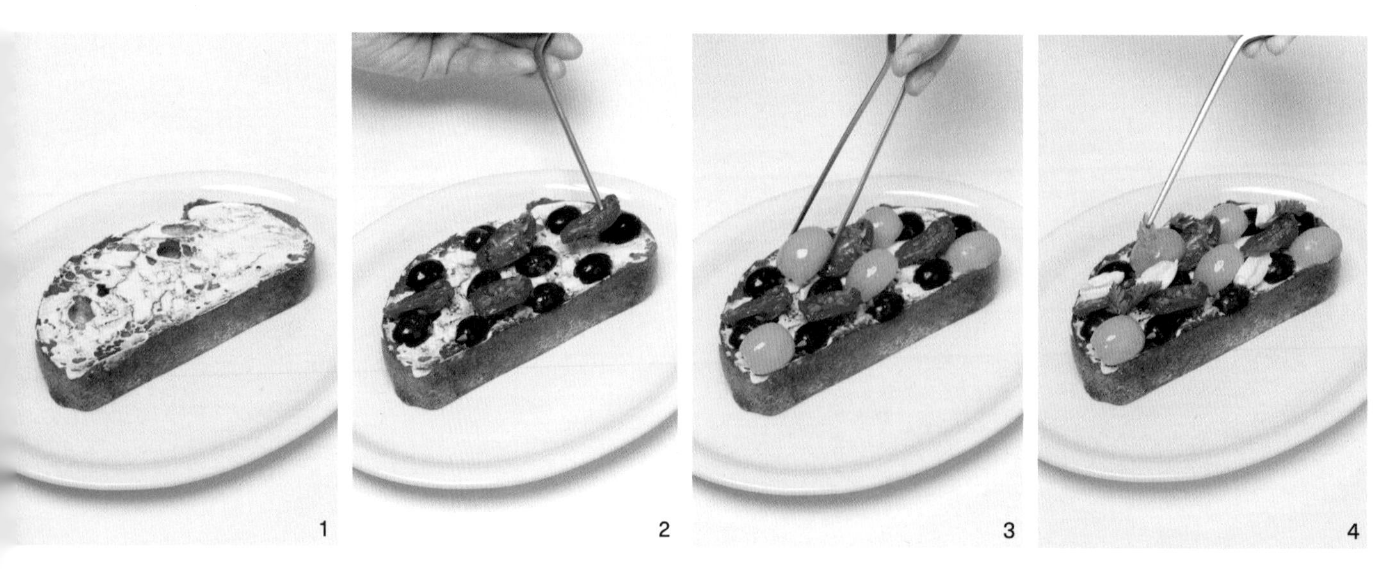

VEGETABLE CURRY WITH LENTILS

채소 커리와 렌틸콩

남은 채소를 활용해 만든 커리인 만큼 어떤 채소를 곁들여도 잘 어울린답니다. 영양분이 풍부하고 칼로리가 낮은 렌틸콩을 곁들여 건강도 함께 챙겨보세요.

As a curry made with vegetables strained from stock, it pairs well with any variety of vegetables. Serve it with nutritious, low-calorie lentils for a healthier option.

1그릇 분량
Makes 1 dish

INGREDIENTS

채소 커리와 렌틸콩

채소 커리(19p)	80g
카놀라유	약간
애호박	1/8개
새송이버섯	1/2개
레드파프리카	1/4개
청경채	1/2개
얇은 아스파라거스	3줄
소금	약간
후추	약간
렌틸콩(통조림)	50g

Vegetable Curry with Lentils

80 g	vegetable curry (p.19)
Q.S.	canola oil (rapeseed oil)
1/8	zucchini
1/2	king oyster mushroom
1/4	red bell pepper
1/2	bok choy
3	thin asparagus spears
Q.S.	salt
Q.S.	ground black pepper
50 g	canned lentils

HOW TO MAKE

채소 커리와 렌틸콩

1. 채소 커리를 뜨겁게 가열한다.
2. 팬에 카놀라유를 두르고 애호박, 새송이, 레드파프리카, 청경채, 얇은 아스파라거스를 구운 후 소금, 후추로 간한다.
3. 팬에 카놀라유를 두르고 렌틸콩을 볶은 후 소금, 후추로 간한다.

Vegetable Curry with Lentils

1. Heat the vegetable curry until hot.
2. Pan-fry the zucchini, king oyster mushroom, red bell pepper, bok choy, and asparagus in canola oil. Season with salt and black pepper.
3. Add extra canola oil to the pan, cook the lentils, then season with salt and black pepper.

PLATING

1. 그릇에 따듯한 채소 커리를 담는다.
2. 그릇 중앙에 볶은 렌틸콩을 올린다.
3. 구운 채소를 조화롭게 올려 마무리한다.

1. Ladle the warm vegetable curry onto a plate.
2. Place the cooked lentils in the center.
3. Arrange the pan-fried vegetables in a balanced presentation.

1

2

3

09

CURRY-COATED ROASTED CAULIFLOWER

커리 소스를 발라 구운 콜리플라워

콜리플라워에 커리 소스를 발라 오븐에 한번, 직화로 또 한번 구워내 불맛이 매력적인 메인 요리입니다.

This main dish showcases curry-coated cauliflower, roasted in the oven and lightly torched for a captivating smoky char.

1그릇 분량
Makes 1 dish

INGREDIENTS

커리 소스를 발라 구운 콜리플라워

채소 커리(19p)	50g
콜리플라워	1/2개
식용유	약간

가니쉬

식용유	약간
사과	1/4개
래디시	2개
적근대	2장
소금	약간
후추	약간
엑스트라 버진 올리브오일	약간
셰리 와인 비네거	약간

Curry-Coated Roasted Cauliflower

50 g	vegetable curry (p.19)
1/2	cauliflower
Q.S.	vegetable oil

Garnishes

Q.S.	vegetable oil
1/4	apple
2	radishes
2	red Swiss chard leaves
Q.S.	salt
Q.S.	ground black pepper
Q.S.	extra virgin olive oil
Q.S.	sherry wine vinegar

Curry-Coated Roasted Cauliflower

HOW TO MAKE

커리 소스를 발라 구운 콜리플라워

1. 채소 커리를 뜨겁게 가열한다.
2. 콜리플라워는 양옆을 평평하게 다듬어 성형한다.
3. 팬에 식용유를 두르고 약불에서 콜리플라워 내부까지 고루 익힌다.
4. 콜리플라워 단면에 스패출러로 채소 커리를 바른다.
5. 180°C로 예열된 오븐에 **4**를 넣고 2분간 굽는다.
6. 오븐에 구운 콜리플라워 단면을 토치로 그을려 불맛을 낸다.

Curry-Coated Roasted Cauliflower

1. Heat the vegetable curry until hot.
2. Trim the cauliflower to flatten both sides.
3. Pan-fry the cauliflower in vegetable oil over low heat until evenly cooked through.
4. Coat the trimmed top with the vegetable curry using a spatula.
5. Roast (**4**) in a preheated 356°F (180°C) oven for 2 minutes.
6. Blowtorch the top of the oven-roasted cauliflower for a smoky char.

가니쉬

1. 팬에 식용유를 두르고 한입 크기로 썬 래디시와 사과, 적근대를 살짝 볶는다.
2. **1**을 트레이에 옮겨 담고 소금, 후추, 올리브오일, 셰리 와인 비네거로 버무린다.

Garnishes

1. Gently pan-fry bite-sized pieces of apple, radish, and red Swiss chard in vegetable oil.
2. Transfer (**1**) to a tray, then gently dress with salt, black pepper, olive oil, and sherry wine vinegar.

PLATING

1. 그릇에 커리 소스를 발라 구운 콜리플라워를 담는다.
2. 볶은 채소와 과일을 곁들여 마무리한다.

1. Place the curry-coated roasted cauliflower on a plate.
2. Serve with the dressed vegetables and fruit on the side.

NAMUL PESTO WITH GLUTEN-FREE PASTA

나물 페스토와 글루텐 프리 파스타

앞서 만든 나물 페스토를 응용해서 만든 파스타입니다. 밀가루 음식이 잘 소화되지 않아 부담스럽다면 글루텐 프리 파스타로 만들어 보세요.

This pasta dish is made using a variation of previously prepped namul pesto. Opt for gluten-free pasta if you have difficulty digesting flour-based foods.

1그릇 분량
Makes 1 dish

INGREDIENTS

나물 페스토와 글루텐 프리 파스타

건조 글루텐 프리 푸실리	80g
나물 페스토(23p)	30g
엑스트라 버진 올리브오일	약간
소금	약간
후추	약간
바질	20g
셰리 와인 비네거	약간

플레이팅

말린 토마토 & 올리브 절임(47p)	약간
바질	5g
구운 캐슈넛	2알

Namul Pesto with Gluten-Free Pasta

80 g	dry gluten-free fusilli
30 g	namul pesto (p.23)
Q.S.	extra virgin olive oil
Q.S.	salt
Q.S.	ground black pepper
20 g	fresh basil leaves
Q.S.	sherry wine vinegar

Plating

Q.S.	dried tomato & olive preserves (p.47)
5 g	fresh basil leaves
2	roasted cashew nuts

1

2

3

HOW TO MAKE

나물 페스토와 글루텐 프리 파스타

1. 소금물에 글루텐 프리 푸실리를 넣고 삶는다.

TIP. 조리 시간은 제품마다 다르므로 포장지에 적힌 시간을 따른다.
소금물은 물 1L, 소금 10g 비율로 끓여 사용한다.

2. 믹싱볼에 **1**을 담은 후 나물 페스토, 올리브오일, 소금, 후추를 넣고 고루 버무린다.

3. 다른 믹싱볼에 바질과 셰리 와인 비네거, 올리브오일, 소금, 후추를 넣고 가볍게 버무린다.

TIP. 바질이 없으면 루콜라 등 다른 잎채소로 대체 가능하다.

Namul Pesto with Gluten-Free Pasta

1. Cook the gluten-free fusilli in salted boiling water.

TIP. Cooking times may vary by brand, so refer to the package instructions.
For the salted water, use a ratio of 10 g of salt per 1 L of water and bring to a boil.

2. Transfer (**1**) to a mixing bowl, then toss evenly with the namul pesto, olive oil, salt, and black pepper.

3. In a separate mixing bowl, lightly dress the basil leaves with the sherry wine vinegar, olive oil, salt, and black pepper.

TIP. If basil is unavailable, other leafy greens like arugula can be used as substitutes.

PLATING

1. 그릇에 나물 페스토에 버무린 글루텐 프리 파스타를 담는다.
2. 말린 토마토 & 올리브 절임과 바질을 올린다.
3. 구운 캐슈넛을 갈아 올려 마무리한다.

1. Place the cooked gluten-free pasta tossed with namul pesto.
2. Top with the dried tomato & olive preserves, along with the dressed basil leaves.
3. Grate the roasted cashew nuts on top.

1

2

3

MUSHROOM-FLAVORED ORECCHIETTE

버섯 풍미의 오레키에테

수분감이 풍부한 버섯과 쫄깃한 오레키에테가 만나 재미있는 식감이 돋보이는 파스타입니다. 숙성된 발사믹 식초가 맛의 레이어를 더하며 복합적인 풍미를 선사합니다.

This pasta showcases an intriguing mix of textures, pairing juicy mushrooms with chewy orecchiette. Aged balsamic vinegar adds layered flavors, bringing complexity to the dish.

1그릇 분량
Makes 1 dish

INGREDIENTS

버섯 풍미의 오레키에테

로즈메리	1줄
만가닥버섯	20g
양송이버섯	2개
애느타리버섯	20g
생표고버섯	1개
건조 오레키에테	80g
엑스트라 버진 올리브오일	약간
소금	약간
후추	약간
다진 마늘	20g
채수(15p)	200g

플레이팅

와일드 루콜라	10g
발사믹 식초	약간

* DOP(Denominazione di Origine Protetta) 혹은 IGP(Indicazione Geografica Protetta) 인증을 받은 발사믹 식초를 사용한다.

Mushroom-Flavored Orecchiette

1	fresh rosemary sprig
20 g	beech mushroom
2	button mushrooms
20 g	oyster mushroom
1	fresh shiitake mushroom
80 g	dry orecchiette
Q.S.	extra virgin olive oil
Q.S.	salt
Q.S.	ground black pepper
20 g	chopped garlic
200 g	vegetable stock (p.15)

Plating

10 g	fresh wild arugula
Q.S.	balsamic vinegar

* Use DOP (Denominazione di Origine Protetta) or IGP (Indicazione Geografica Protetta)-certified balsamic vinegar.

HOW TO MAKE

버섯 풍미의 오레키에테

1. 로즈메리를 최대한 곱게 다진다.

TIP. 가을, 겨울철처럼 억센 상태의 줄기가 아니면 잎과 줄기를 모두 다져 사용한다.

2. 만가닥버섯, 양송이버섯, 애느타리버섯, 생표고버섯을 한입 크기로 자른다.
3. 소금물에 오레키에테를 넣고 삶는다.

TIP. 조리 시간은 제품마다 다르므로 포장지에 적힌 시간을 따른다. 소금물은 물 1L, 소금 10g 비율로 끓여 사용한다.

4. 팬에 올리브오일을 넉넉히 두른 후 **2**를 약불에서 볶는다.
5. 버섯의 수분이 모두 날아가고 갈색빛이 돌면 소금, 후추로 간한다.
6. **5**가 담긴 팬을 살짝 기울인 후, 한쪽에 올리브오일과 다진 마늘을 넣고 약불에서 타지 않게 볶는다.
7. 마늘이 어느 정도 익으면 **5**와 함께 볶다가 삶은 오레키에테, 채수를 넣고 에멀전한다.
8. 에멀젼 상태(물과 기름이 유화된 상태)가 되면 소금, 후추로 간한다.

Mushroom-Flavored Orecchiette

1. Chop the rosemary as finely as possible.

TIP. Use the entire rosemary, including the leaves and stem, unless the stem is too tough, as it often is in autumn and winter.

2. Cut the beech, button, shiitake and oyster mushrooms into bite-sized pieces.
3. Cook the orecchiette in a pot of salted boiling water.

TIP. Cooking times vary by brand, so refer to the package instructions.
For the salted water, use a ratio of 10 g of salt per 1 L of water and bring to a boil.

4. In a pan with ample vegetable oil, cook (**2**) over low heat.
5. Once the moisture evaporates from the mushrooms and they turn brown, season with salt and black pepper.
6. Slightly tilt the pan and cook the chopped garlic in olive oil on the empty side over low heat, keeping it from burning.
7. Once the garlic is cooked, combine it with (**5**). Add the cooked orecchiette and vegetable stock, then emulsify.
8. Once emulsified, season with salt and black pepper.

PLATING

1. 그릇에 각종 버섯과 오레키에테를 담는다.
2. 와일드 루콜라를 올린다.
3. 발사믹 식초를 뿌려 마무리한다.

1. Arrange the assorted mushrooms and orecchiette on a plate.
2. Top with the wild arugula.
3. Finish with a drizzle of balsamic vinegar.

1

2

3

12

TORTELLINI STUFFED WITH MUSHROOM DUXELLES & VEGETABLE CONSOMMÉ

양송이 뒥셀을 채운 토르텔리니와 채소 콩소메

앞서 만든 양송이 뒥셀을 소로 채워 만두처럼 빚어낸 토르텔리니 파스타입니다. 따듯한 채소 콩소메를 곁들여 만둣국처럼 즐겨보세요.

This pasta, tortellini, is filled with previously prepped mushroom duxelles before being shaped into a dumpling. Enjoy it as a comforting dumpling soup served with warm vegetable consommé.

1그릇 분량
Makes 1 dish

INGREDIENTS

양송이 뒥셀을 채운 토르텔리니와 채소 콩소메

만두피	10장
양송이 뒥셀(43p)	50g
채수(15p)	400g
건조 포르치니버섯	1조각
소금	약간

Tortellini Stuffed with Mushroom Duxelles & Vegetable Consommé

10	dumpling wrappers
50 g	mushroom duxelles (p.43)
400 g	vegetable stock (p.15)
1	dried porcini mushroom slice
Q.S.	salt

플레이팅

처빌	10g
후추	약간

Plating

10 g	fresh chervil
Q.S.	ground black pepper

1
2
3
4
5

HOW TO MAKE

양송이 뒥셀을 채운 토르텔리니와 채소 콩소메

1. 만두피를 원형 커터(∅6cm)로 자른다.
2. 양송이 뒥셀을 파이핑백에 담아 만두피 중앙에 파이핑한다.
3. 만두 모양으로 빚어 토르텔리니를 만든다.

TIP. 만두피가 마르면 물을 뿌려 촉촉한 상태를 유지한다. 완성된 토르텔리니는 가위로 가장자리를 깔끔하게 정리한다.

4. 냄비에 채수와 건조 포르치니버섯을 넣고 약불로 끓인 후 소금으로 간한다.
5. 다른 냄비에 물을 끓인 후 **3**을 넣고 30~50초간 삶는다.

Tortellini Stuffed with Mushroom Duxelles & Vegetable Consommé

1. Cut the dumpling wrappers with a 6 cm round cutter.
2. Transfer the mushroom duxelles to a piping bag and pipe a portion into the center of each wrapper.
3. Fold each wrapper and shape it into a dumpling-like form to make tortellini.

**TIP. If the wrappers dry out, spray with water to keep them moist.
For a clean finish, trim the edges of the tortellini with scissors.**

4. In a pot, simmer the vegetable stock with a slice of dried porcini mushroom over low heat, then season with salt.
5. In a separate pot of boiling water, cook (**3**) for 30 to 50 seconds.

PLATING

1. 그릇에 포르치니 향이 우러난 채소 콩소메를 담는다.
2. 삶은 토르텔리니를 담는다.
3. 처빌을 올리고 후추를 뿌려 마무리한다.

TIP. **처빌은 찬물에 담가 싱싱하게 만들어 사용한다.**

1. Ladle the porcini-infused vegetable consommé into a bowl.
2. Place the cooked tortellini into the consommé.
3. Finish with the chervil and ground black pepper.

TIP. **Soak the chervil in ice water to refresh it before use.**

1 2 3

13

HERB-INFUSED SUNCHOKE WITH SOY MILK MAYONNAISE

허브 향의 돼지감자와 두유 마요네즈

돼지감자에 허브 향을 입혀 구워내 산뜻하면서도 포만감을 주는 요리입니다. 앞서 만든 두유 마요네즈를 함께 곁들이면 간단하면서도 고급스러운 요리가 완성됩니다.

This herb-infused sunchoke dish is cooked to be refreshing while still satisfying. Serve it with previously prepped soy milk mayonnaise to complete a simple yet refined dish.

1그릇 분량
Makes 1 dish

INGREDIENTS

허브 향의 돼지감자와 두유 마요네즈

돼지감자	200g
중력분	약간
식용유	약간
마늘	2쪽
타임	5g
로즈메리	5g
소금	약간
후추	약간

Herb-Infused Sunchoke with Soy Milk Mayonnaise

200 g	sunchokes
Q.S.	all-purpose flour
Q.S.	vegetable oil
2	garlic cloves
5 g	fresh thyme
5 g	fresh rosemary
Q.S.	salt
Q.S.	ground black pepper

플레이팅

말린 토마토 & 올리브 절임(47p)	약간
레몬	1/6조각
두유 마요네즈(27p)	30g

Plating

Q.S.	dried tomato & olive preserves (p.47)
1/6	lemon
30 g	soy milk mayonnaise (p.27)

HOW TO MAKE

허브 향의 돼지감자와 두유 마요네즈

1. 끓는 물에 돼지감자를 넣고 내부가 완전히 익을 때까지 삶는다.

TIP. 돼지감자는 흐르는 물에 칫솔로 구석구석 씻어 사용한다. 젓가락으로 찔렀을 때 부드럽게 들어갈 정도로 푹 익힌다.

2. 돼지감자는 충분히 식힌 후 중력분을 고루 묻힌다.

TIP. 밀가루는 최대한 얇게 묻힌다.

3. 중불로 예열한 냄비에 식용유를 넉넉히 두르고 2를 튀겨내듯 익힌다.

4. 전체적으로 황금빛이 돌면 약불로 줄인 후, 손으로 으깬 마늘과 타임, 로즈메리를 넣는다.

TIP. 기름에 마늘과 허브 향이 우러나면 돼지감자에 기름을 끼얹어가며 향을 더한다.

5. 돼지감자를 건져 믹싱볼에 옮겨 담은 후 소금, 후추로 간한다.

Herb-Infused Sunchoke with Soy Milk Mayonnaise

1. Cook the sunchokes in boiling water until fully cooked through.

TIP. Scrub the sunchokes thoroughly with a toothbrush under running water.
Cook until a chopstick easily pierces through.

2. Once cooled, coat the sunchokes evenly with all-purpose flour.

TIP. Apply the flour as lightly as possible.

3. In a preheated pot over medium heat, cook (2) with ample vegetable oil, as if frying.

4. When they turn golden brown, reduce the heat to low. Add crushed garlic, thyme, and rosemary.

TIP. Once the garlic and herbs become aromatic in the oil, baste the sunchokes in it to infuse them with fragrance.

5. Remove the sunchokes from the pot, transfer them to a mixing bowl, then season with salt and black pepper.

PLATING

1. 그릇에 돼지감자를 담는다.
2. **1** 위에 말린 토마토 & 올리브 절임을 올린다.
3. 토치로 그을린 레몬을 곁들이고, 두유 마요네즈를 소스 그릇에 담아 함께 제공한다.

1. Arrange the cooked sunchokes on a plate.
2. Garnish with the dried tomato & olive preserves on (**1**).
3. Serve with a blowtorched lemon wedge on the plate and the soy milk mayonnaise in a sauce bowl.

1

2

3

MUSHROOM DUXELLES & PERILLA LEAF ROLL CUTLETS

양송이 뒥셀 깻잎말이 커틀릿

양송이 뒥셀을 깻잎으로 돌돌 말아서 빵가루 옷을 입혀 바삭하게 튀겨낸 요리입니다. 파프리카 마요네즈에 찍어 안주나 간식으로 가볍게 즐겨보세요.

This bite features mushroom duxelles wrapped in perilla leaves, breaded, and fried until crispy. Enjoy it as a light snack or a side dish with drinks, served with paprika mayonnaise for dipping.

10조각 분량

Makes 10 pieces

INGREDIENTS

튀김 반죽 ●

박력분	250g
식수	300g
소금	7g

Batter ●

250 g	cake flour
300 g	drinking water
7 g	salt

양송이 뒥셀 깻잎말이 커틀릿

깻잎	10장
양송이 뒥셀(43p)	100g
튀김 반죽 ●	전량
건식 빵가루	300g
식용유	약간

Mushroom Duxelles & Perilla Leaf Roll Cutlets

10	fresh perilla leaves
100 g	mushroom duxelles (p.43)
Whole batter mixture ●	
300 g	dry breadcrumbs
Q.S.	vegetable oil

플레이팅

이탈리안 파슬리	10g
두유 마요네즈(27p)	100g
스모크파프리카 파우더	15g

Plating

10 g	fresh Italian parsley
100 g	soy milk mayonnaise (p.27)
15 g	smoked paprika powder

Batter

Mushroom Duxelles & Perilla Leaf Roll Cutlets

1

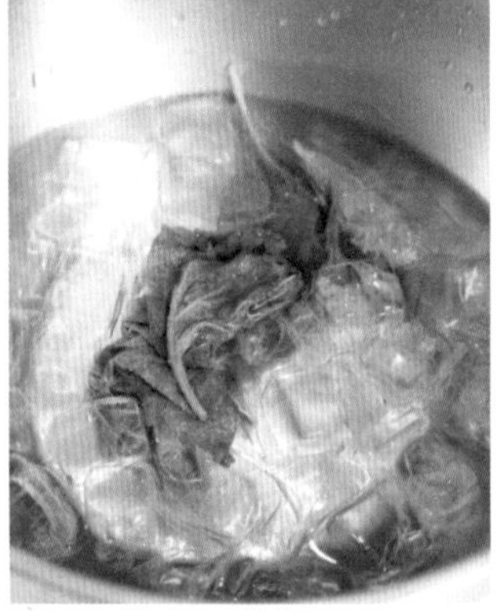

2

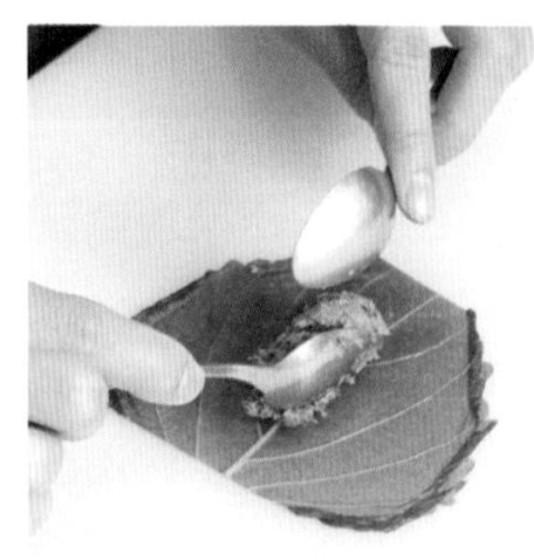

3

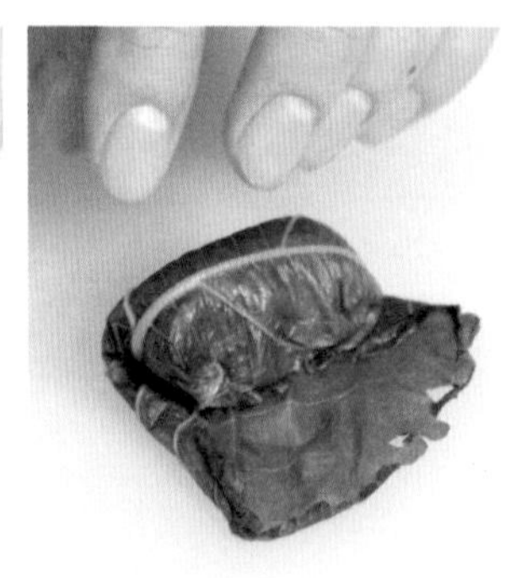

4

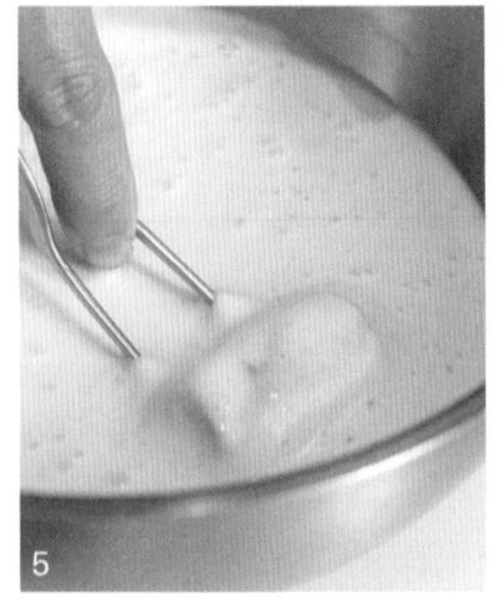

5

6

HOW TO MAKE

튀김 반죽

믹싱볼에 모든 재료를 담고 고루 섞어 준비한다.

Batter

Combine all the ingredients in a mixing bowl and mix thoroughly.

양송이 뒥셀 깻잎말이 커틀릿

1. 깻잎은 끓는 물에 데친 후 얼음물에 식힌다.
2. 물기를 제거한 **1**을 넓게 펼친 후 양쪽 가장자리를 자른다.
3. 자른 깻잎 위에 양송이 뒥셀을 적당량 올린다.
4. 깻잎의 양쪽 끝을 접고 돌돌 말아 원통 모양으로 만든다.

TIP. **튀어나온 부분은 깔끔하게 정리한다.**

5. 튀김 반죽에 **4**를 담갔다가 건진 후 건식 빵가루를 고루 묻힌다.
6. 180°C로 예열한 식용유에 **5**를 넣고 1~2분간 튀긴다.

Mushroom Duxelles & Perilla Leaf Roll Cutlets

1. Blanch the perilla leaves in boiling water, then transfer them to ice water to cool.
2. Unfold each drained perilla leaf and trim both edges.
3. Place a portion of mushroom duxelles on each leaf.
4. Fold in the edges and roll each perilla leaf into a cylinder shape.

TIP. **For a clean finish, trim any overhanging parts.**

5. Dip (**4**) into the batter, lift it out, and coat it evenly with dry breadcrumbs.
6. Fry (**5**) in vegetable oil preheated to 356°F (180°C) for 1 to 2 minutes.

—

PLATING

1. 이탈리안 파슬리를 잘게 다진다.
2. 두유 마요네즈와 스모크파프리카 파우더를 섞어 소스를 만든다.
3. **2**를 소스 그릇에 옮겨 담고 다진 파슬리를 올린 후,
   양송이 뒥셀 깻잎말이 커틀릿과 함께 제공한다.

1. Finely chop the Italian parsley.
2. Mix the soy milk mayonnaise with smoked paprika powder to make the sauce.
3. Transfer (**2**) to a sauce bowl and sprinkle the chopped parsley on top.
   Serve alongside the mushroom duxelles & perilla leaf roll cutlets.

1

2

3

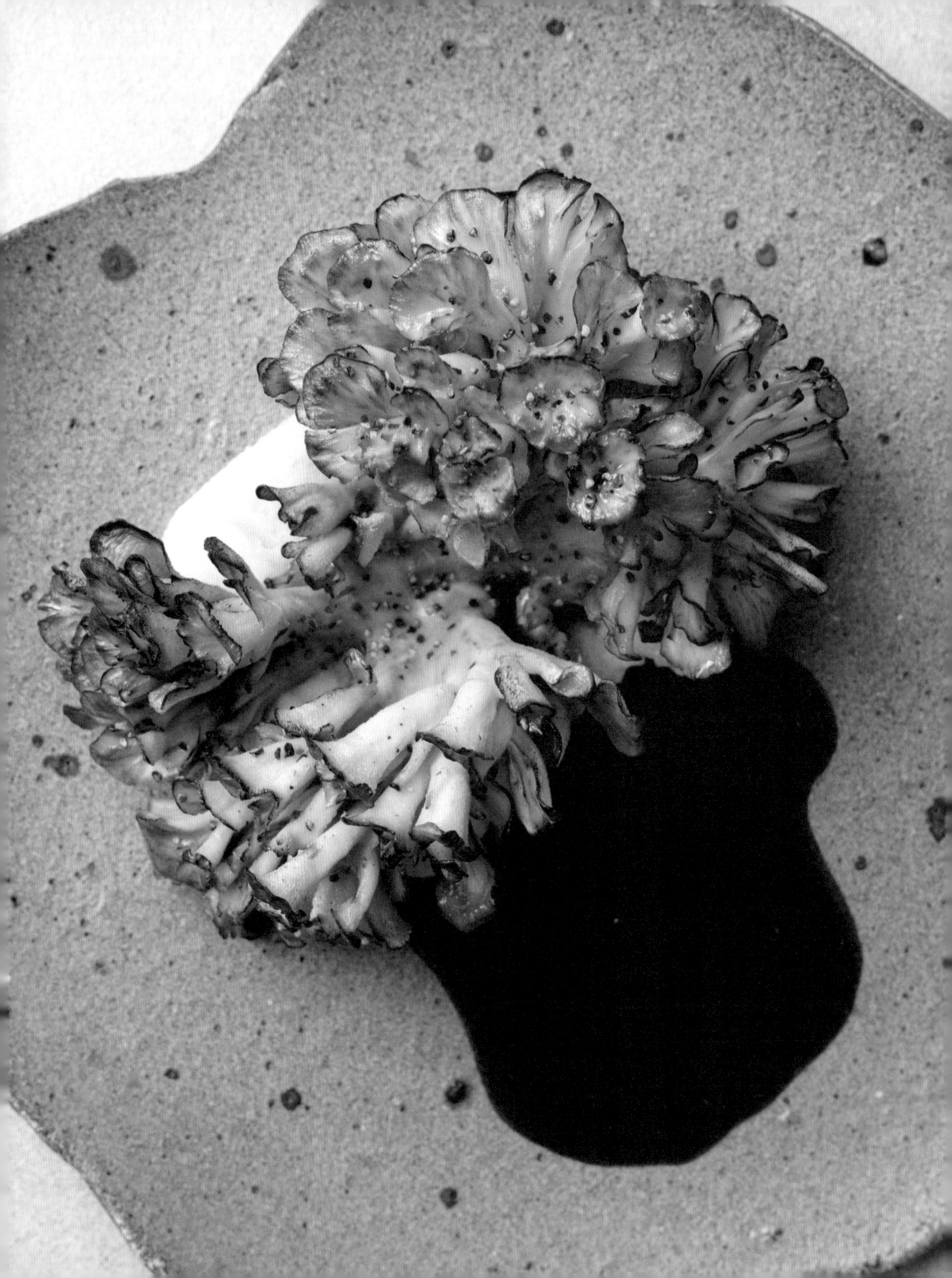

MAITAKE MUSHROOM STEAK WITH BALSAMIC & VEGETABLE REDUCTION SAUCE

잎새버섯 스테이크와 발사믹 & 채소 리덕션 소스

잎새버섯을 스테이크처럼 통째로 구워 수분감을 보존하고 감칠맛을 한껏 끌어낸 요리입니다. 채수와 발사믹 식초를 졸여 만든 소스와 체에 곱게 내린 감자를 함께 곁들여 무게감을 더해줍니다.

This dish showcases a whole-roasted maitake mushroom, cooked like a steak to retain its juicy texture and deepen its savory flavor. A rich reduction sauce made from vegetable stock and balsamic vinegar, served with finely sieved potatoes, adds satisfying depth and richness to the dish.

1그릇 분량
Makes 1 dish

INGREDIENTS

잎새버섯 스테이크와 발사믹 & 채소 리덕션 소스

채수(15p)	500g
잎새버섯	200g
엑스트라 버진 올리브오일	약간
발사믹 식초	1작은술
소금	약간
후추	약간
감자	1개

* DOP 혹은 IGP 인증을 받은 발사믹 식초를 사용한다.

Maitake Mushroom Steak with Balsamic & Vegetable Reduction Sauce

500 g	vegetable stock (p.15)
200 g	maitake mushroom
Q.S.	extra virgin olive oil
1 ts	balsamic vinegar
Q.S.	salt
Q.S.	ground black pepper
1	potato

* Use DOP or IGP-certified balsamic vinegar.

1
2
3
4
5
6
7

HOW TO MAKE

잎새버섯 스테이크와 발사믹 & 채소 리덕션 소스

1. 냄비에 채수를 넣고 약 1/6이 될 때까지 졸여 맛을 응축시킨다.
2. 1에 발사믹 식초와 올리브오일을 넣고 약불로 가열해 리덕션 소스를 만든다.
3. 잎새버섯 겉면에 올리브오일을 바른 후 약 200°C로 예열된 오븐에 3~5분간 굽는다.
4. 구운 잎새버섯 겉면을 토치로 그을려 불맛을 내고 소금, 후추로 간한다.
5. 감자를 끓는 물에 껍질째 삶아 완전히 익힌다.

TIP. **젓가락으로 찔렀을 때 부드럽게 들어갈 정도로 푹 익힌다.**

6. 부드럽게 익은 감자의 껍질을 제거하고 체에 곱게 내린다.
7. 6에 소금, 후추, 올리브오일을 넣고 고루 섞는다.

Maitake Mushroom Steak with Balsamic & Vegetable Reduction Sauce

1. Pour the vegetable stock into a pot and reduce it by one-sixth to concentrate its flavor.
2. Add the balsamic vinegar and olive oil to (**1**), then cook over low heat to make the reduction sauce.
3. Brush the mushroom skin with olive oil and roast in an oven preheated to 392°F (200°C) for 3 to 5 minutes.
4. Blowtorch the roasted mushroom for a smoky char, then season with salt and black pepper.
5. Boil the potatoes with their skin on in boiling water until fully cooked through.

TIP. **Cook until a chopstick easily pierces through.**

6. Peel the cooked potatoes and press them through a fine-mesh sieve.
7. Add salt, black pepper, and olive oil to (**6**), then mix thoroughly.

PLATING

1. 체에 곱게 내린 감자를 그릇에 담는다.
2. 감자 위에 구운 잎새버섯 스테이크를 올린다.
3. 발사믹 & 채소 리덕션 소스를 곁들여 마무리한다.

1. Spoon the finely sieved potatoes onto a plate.
2. Place the roasted maitake mushroom steak on the potatoes.
3. Serve with the balsamic & vegetable reduction sauce.

1

2

16

MINI PAPRIKA STUFFED WITH VEGETABLES AND MUSHROOMS

채소와 버섯으로 속을 채운 미니 파프리카

미니 파프리카를 태워 껍질과 씨앗을 제거한 후 그 속을 각종 채소로 채운 요리입니다. 단맛이 증폭된 파프리카와 발사믹 식초 베이스의 새콤한 소스가 조화롭게 어우러집니다.

This dish features mini paprikas stuffed with assorted vegetables, with their charred skin and seeds removed. The amplified sweetness of the blowtorched paprika pairs beautifully with a tangy balsamic-based sauce, creating a harmonious balance.

1그릇 분량
Makes 1 dish

INGREDIENTS

채소와 버섯으로 속을 채운 미니 파프리카

미니 파프리카(레드)	4개
양배추	100g
빵가루	10g
볶은 캐슈넛	5알
식용유	약간
소금	약간
후추	약간
양송이 뒥셀(43p)	100g
마늘	2쪽
샬롯	1개
엑스트라 버진 올리브오일	약간
발사믹 식초	약간

* DOP 혹은 IGP 인증을 받은 발사믹 식초를 사용한다.

Mini Paprika Stuffed with Vegetables and Mushrooms

4	mini paprikas (red)
100 g	cabbage
10 g	breadcrumbs
5	roasted cashew nuts
Q.S.	vegetable oil
Q.S.	salt
Q.S.	ground black pepper
100 g	mushroom duxelles (p.43)
2	garlic cloves
1	shallot
Q.S.	extra virgin olive oil
Q.S.	balsamic vinegar

* Use DOP or IGP-certified balsamic vinegar.

플레이팅

미니고수	약간

Plating

Q.S.	fresh mini cilantro

1
2
3
4
5
6
7
8
9
10
11
12

HOW TO MAKE

채소와 버섯으로 속을 채운 미니 파프리카

1. 미니 파프리카를 껍질째 돌려가며 토치로 까맣게 태운다.

TIP. **토치가 없다면 가스 불을 사용해도 무방하다. 파프리카 껍질을 태우면 단맛이 올라온다.**

2. 검게 탄 파프리카 껍질을 날붙이로 긁어 제거한다.
3. 파프리카의 꼭지 부분을 잘라 안쪽에 있는 씨앗을 제거한다.
4. 양배추를 아주 얇게 채 썬다.
5. 팬에 빵가루를 넣고 갈색빛이 돌 때까지 굽는다.
6. 볶은 캐슈넛은 칼로 듬성듬성 다진다.
7. 식용유를 두른 팬에 **4**를 넣고 센 불에서 빠르게 볶아 수분을 날린 다음 소금, 후추로 간한다.
8. 믹싱볼에 양송이 뒥셀과 **5**, **6**, **7**을 넣어 고루 섞은 후 소금, 후추로 간하여 소를 만든다.
9. **8**을 파이핑백에 옮겨 담아 파프리카 안쪽에 가득 채운 다음 먹기 좋은 크기로 자른다.
10. 마늘과 샬롯은 아주 곱게 다진다.
11. 약불로 예열한 팬에 올리브오일을 넉넉히 두른 후 **10**을 넣고 볶는다.
12. 향이 우러나면 불을 끄고 **11**에 소금, 후추, 발사믹 식초를 넣어 소스를 만든다.

Mini Paprika Stuffed with Vegetables and Mushrooms

1. Char each mini paprika with a blowtorch, rotating it with the skin on.

TIP. **Use a gas flame if a torch is unavailable. Charring the skin brings out the paprika's natural sweetness.**

2. Scrap off the charred black skin with a stainless-steel spatula.
3. Remove the top of each paprika, then scoop out the seeds.
4. Thinly slice the cabbage.
5. Toast the breadcrumbs in a pan until golden brown.
6. Coarsely chop the roasted cashew nuts.
7. Sauté (**4**) in a pan with vegetable oil over high heat to evaporate moisture, then season with salt and black pepper.
8. In a mixing bowl, combine the mushroom duxelles, (**5**), (**6**), and (**7**), then mix thoroughly. Season with salt and black pepper to make the filling.
9. Transfer (**8**) to a piping bag, and pipe it into each paprika until fully stuffed. Slice into bite-sized pieces for easier eating.
10. Finely chop the garlic and shallot.
11. In a preheated pan over low heat, cook (**10**) with ample olive oil.
12. Once aromatic, remove from heat, then stir in salt, black pepper, and balsamic vinegar to make the sauce.

PLATING

1. 그릇에 소스를 담는다.
2. 먹기 좋은 사이즈로 자른 파프리카를 담는다.
3. 미니고수 잎을 올려 마무리한다.

TIP. **미니고수는 찬물에 담가 싱싱하게 만들어 사용한다.**

1. Spoon the sauce onto a plate.
2. Arrange the bite-sized pieces of mini paprikas.
3. Garnish with the mini cilantro.

TIP. **Soak the mini cilantro in ice water to refresh it before use.**

1

2

3

DESSERTS

01

ROASTED BANANA WITH SOY MILK YOGURT

구운 바나나와 두유 요거트

잘 후숙된 바나나를 껍질째 구워 감미로운 맛과 촉촉한 식감이 돋보이는 디저트입니다. 고소한 헤이즐넛과 메이플 시럽, 두유 요거트를 곁들여 드셔보세요.

This dessert showcases a perfectly ripened banana, charred in its peel to deepen its sweetness and enhance its moist texture. Serve it with savory hazelnuts, maple syrup, and soy milk yogurt.

1그릇 분량

Makes 1 dish

INGREDIENTS

구운 바나나

바나나	1개
식용유	약간
헤이즐넛	10알
소금	약간

Roasted Banana

1	banana
Q.S.	vegetable oil
10	hazelnuts
Q.S.	salt

플레이팅

메이플 시럽	30g
타임	5g
소금	약간
두유 요거트(31p)	30g

Plating

30 g	maple syrup
5 g	fresh thyme
Q.S.	salt
30 g	soy milk yogurt (p.31)

1
2
3
4
5

HOW TO MAKE

구운 바나나

1. 바나나를 껍질째 돌려가며 토치로 까맣게 태운다.

TIP. 토치가 없다면 가스 불을 사용해도 무방하다.
구운 후 열이 내부까지 전달되도록 잠시 휴지시킨다.

2. 팬에 식용유를 두르고 헤이즐넛을 노릇하게 굽는다.
3. 2를 칼로 이등분한 후 소금으로 간한다.
4. 바나나 껍질 윗면에 칼집을 넣어 위 껍질을 제거한다.
5. 바나나 과육은 한입 크기로 자른다.

Roasted Banana

1. Char the banana with a blowtorch, rotating it with the peel on.

TIP. Use a gas flame if a torch is unavailable.
Let it rest after charring to allow the heat to penetrate.

2. Toast the hazelnuts in a pan with vegetable oil until golden brown.
3. Halve (2) with a knife and season with salt.
4. Make a slit at the top of the banana peel to remove the top section.
5. Cut the banana flesh into bite-sized slices.

PLATING

1. 그릇에 바나나를 껍질째 담고 메이플 시럽을 뿌린다.
2. **1**에 반으로 자른 헤이즐넛과 타임 잎을 올리고 약간의 소금을 뿌린다.
3. 두유 요거트를 소스 그릇에 담아 함께 제공한다.

1. Place the sliced banana on a plate, with the bottom peel left on, then drizzle with maple syrup.
2. Garnish with the halved hazelnuts, thyme leaves, and a pinch of salt over (**1**).
3. Serve with the soy milk yogurt in a sauce bowl.

1

2

3

CHAMOMILE-INFUSED PEAR WITH GRANOLA & SOY MILK YOGURT

캐모마일 배 절임과 그래놀라 & 두유 요거트

풍부한 과즙과 아삭한 식감이 특징인 한국 배를 캐모마일 시럽에 절여 은은한 꽃 향을 더한 디저트입니다. 바삭한 그래놀라와 두유 요거트를 곁들여 간편하게 즐겨 보세요.

This dessert features a Korean pear, loved for its juicy burst and crisp texture, steeped in chamomile syrup to infuse its delicate floral notes. Serve it with crunchy granola and soy milk yogurt for a simple treat.

1그릇 분량
Makes 1 dish

INGREDIENTS

캐모마일 시럽 •

설탕	480g
물	500g
건조 캐모마일	8g
라임	1개

캐모마일 배 절임
(500g 분량, 보관 기간 냉장 3일)

배	1개
캐모마일 시럽 •	500g

플레이팅

두유 요거트(31p)	200g
비건 그래놀라	20g
브론즈펜넬 잎	약간

Chamomile Syrup •

480 g	sugar
500 g	water
8 g	dried chamomile
1	lime

Chamomile-Infused Pear
(Yields 500 g, and lasts for 3 days when refrigerated)

1	Korean pear
500 g	chamomile syrup •

Plating

200 g	soy milk yogurt (p.31)
20 g	vegan granola
Q.S.	fresh bronze fennel leaves

Chamomile Syrup

Chamomile-Infused Pear

1

2

HOW TO MAKE

캐모마일 시럽

냄비에 설탕, 물, 건조 캐모마일, 라임을 넣고 한소끔 끓여 시럽을 만든다.

Chamomile Syrup

In a pot, combine the dried chamomile, lime, sugar, and water, then bring to a boil to make the syrup.

캐모마일 배 절임

1. 배는 껍질과 씨를 제거하고 먹기 좋은 크기로 자른다.
2. 용기에 **1**을 담은 후 충분히 식힌 캐모마일 시럽을 붓고 밀폐하여 냉장 보관한다.

Chamomile-Infused Pear

1. Peel, core, and cut the Korean pear into bite-sized pieces for easier eating.
2. Place (**1**) in an airtight container, pour over the cooled chamomile syrup, and refrigerate.

PLATING

1. 그릇 아랫부분에 두유 요거트를 담는다.
2. 두유 요거트 위로 캐모마일 배 절임을 담는다.
3. 비건 그래놀라와 브론즈펜넬 잎을 올리고 캐모마일 시럽을 뿌려 마무리한다.

TIP. **브론즈펜넬 잎은 찬물에 담가 싱싱하게 만들어 사용한다.**

1. Spoon the soy milk yogurt into the bottom of a bowl.
2. Arrange the chamomile-infused pear over the soy milk yogurt.
3. Finish with the vegan granola, bronze fennel leaves, and a drizzle of chamomile syrup.

TIP. **Soak the bronze fennel leaves in ice water to refresh them before use.**

1

2

3

03

TOMATO GRANITA WITH BASIL OIL

토마토 그라니타와 바질오일

방울토마토로 만든 시원한 그라니타에 향긋한 바질오일을 곁들여 깔끔하고 상쾌한 맛이 돋보이는 디저트입니다.

Accompanied by fragrant basil oil, this chilled cherry tomato granita highlights a simple, refreshing flavor.

1그릇 분량
Makes 1 dish

INGREDIENTS

토마토 그라니타

대추방울토마토	600g
물	300g
레몬즙	20g
설탕	230g
소금	2g

Tomato Granita

600 g	grape tomatoes
300 g	water
20 g	lemon juice
230 g	sugar
2 g	salt

플레이팅

바질오일	약간
아마란스	약간

* 바질로 그린 허브오일(39p)을 만들어 사용한다.
* 아마란스 대신 적색 허브류나 구하기 쉬운 허브류를 사용해도 좋다.

Plating

Q.S.	basil oil
Q.S.	fresh amaranth

* Make the green herb oil (p.39) with basil.
* Other red or affordable herbs can be used as substitutes for amaranth.

1

2

3

4

HOW TO MAKE

토마토 그라니타

1. 대추방울토마토는 세척 후 꼭지를 제거한다.
2. 믹서에 **1**과 물, 레몬즙, 설탕, 소금을 넣고 곱게 간다.

TIP. **토마토 당도에 따라 설탕의 양을 조절한다.**

3. 완성된 그라니타 베이스는 체에 내려 넓은 밀폐 용기에 옮겨 담은 후 냉동고에서 12시간 이상 얼린다.
4. 그라니타가 충분히 얼면 냉동고에서 꺼낸 후 포크로 긁어 입자를 만든다.

Tomato Granita

1. Rinse the grape tomatoes and remove the stems.
2. In a blender, combine (**1**) with the lemon juice, water, sugar, and salt, then blend until smooth.

TIP. **Adjust the sugar based on the sweetness of tomatoes.**

3. Strain the mixture through a sieve, transfer to a wide, airtight container, then freeze for at least 12 hours.
4. Once frozen, remove from the freezer and scrape with a fork to form granita flakes.

PLATING

1. 그릇에 토마토 그라니타를 담는다.
2. 아마란스를 올린다.

TIP. **아마란스는 찬물에 담가 싱싱하게 만들어 사용한다.**

3. 그라니타에 바질오일을 곁들여 마무리한다.

1. Place the tomato granita in a bowl.
2. Top with the amaranth.

TIP. **Soak the amaranth in ice water to refresh it before use.**

3. Finish with a spoonful of basil oil along the edge of the granita.

1

2

04

VEGETABLE CHIPS

채소 과자

남은 재료를 활용할 수 있는 제로 웨이스트 레시피로 만들어 친환경적일 뿐만 아니라 맛도 좋고 영양도 풍부한 채소 과자입니다.

These vegetable chips, crafted using a zero-waste recipe that makes use of vegetables strained from stock, are both eco-friendly and packed with flavor and nutrients.

1그릇 분량

Makes 1 dish

INGREDIENTS

채소 과자

잔여 채소(17p)	250g
해바라기씨A	75g
물	60g
메이플 시럽	8g
참깨A	5g
불린 아마씨	100g
소금	5g
강력분	54g

토핑

쪽파	50g
엑스트라 버진 올리브오일	약간
소금	약간
후추	약간
해바라기씨B	50g
검은깨	10g
참깨B	10g

플레이팅

나물 페스토(23p)	약간

Vegetable Chips

250 g	cooked vegetable solids (p.17)
75 g	sunflower seeds (A)
60 g	water
8 g	maple syrup
5 g	sesame seeds (A)
100 g	soaked flaxseeds
5 g	salt
54 g	bread flour

Topping

50 g	chives
Q.S.	extra virgin olive oil
Q.S.	salt
Q.S.	ground black pepper
50 g	sunflower seeds (B)
10 g	black sesame seeds
10 g	sesame seeds (B)

Plating

Q.S.	namul pesto (p.23)

1
2
3
4
5
6

HOW TO MAKE

채소 과자

1. 쪽파를 얇게 썰어 올리브오일과 소금, 후추로 버무린다.
2. 믹서에 잔여 채소와 해바라기씨A, 물, 메이플 시럽, 참깨A, 불린 아마씨, 소금을 넣고 곱게 간다.
3. 믹싱볼에 **2**를 옮겨 담고 강력분을 고루 섞어 반죽한다.
4. 내열 실리콘 매트 위로 **3**을 얇고 넓게 펼쳐 팬닝한다.
5. **4** 위로 **1**과 해바라기씨B, 검은깨, 참깨B를 고루 올리고 소금으로 간한다.
6. 130°C로 예열된 오븐에 **5**를 넣고 약 40분간 굽는다.

TIP.
- 수분이 모두 날아가 바삭한 상태가 되면 마무리한다.
- 오븐의 성능에 따라 굽는 시간에 차이가 있을 수 있다.
- 완성된 채소 과자는 실리카겔을 넣은 밀폐 용기나 건조기에 보관해 바삭한 식감을 유지한다.

Vegetable Chips

1. Thinly slice the chives and dress with olive oil, salt, and black pepper.
2. In a blender, combine the cooked vegetable solids, sunflower seeds (A), water, maple syrup, sesame seeds (A), soaked flaxseeds, and salt, then blend until smooth.
3. Transfer (**2**) to a mixing bowl, add the bread flour, and mix until fully incorporated.
4. Spread (**3**) thinly and evenly onto a heat-resistant silicone mat.
5. Evenly sprinkle (**1**), sunflower seeds (B), black sesame seeds, sesame seeds (B), and salt over (**4**).
6. Roast (**5**) in a preheated oven at 266°F (130°C) for about 40 minutes.

TIP.
- Roast until all moisture has evaporated.
- Cooking times may vary depending on oven performance.
- Store the vegetable chips in a dehydrator or an airtight container with silica gel to keep them crispy.

PLATING

1. 완성된 채소 과자는 먹기 좋은 사이즈로 잘라 그릇에 올린다.
2. 나물 페스토는 소스 그릇에 담아 함께 제공한다.

1. Break the vegetable chips into small pieces for easier eating and arrange them on a plate.
2. Serve with the namul pesto in a sauce bowl.

1

2

수상 ACCOLADES

미쉐린 가이드 서울 | 부산 2025 미쉐린 1스타
(The MICHELIN Guide Seoul & Busan 2025, 1 Star)

위아 스마트 그린 가이드 2024 5래디시
(We're Smart Green Guide 2024, 5 Radishes)

2024 서울미식 100선
(100 Taste of Seoul 2024)

2024 서울채식 50선
(50 Plant-Based Restaurant of Seoul 2024)

블루리본 서베이 2025 리본 2개
(Blue Ribbon Survey 2025, 2 Ribbons)

미쉐린 가이드 서울 | 부산 2024
(The MICHELIN Guide Seoul & Busan 2024)

블루리본 서베이 2024
(Blue Ribbon Survey 2024)

2023 서울미식 100선
(100 Taste of Seoul 2023)

2023 서울채식 50선
(50 Plant-Based Restaurant of Seoul 2023)

협업 COLLABORATIONS

2024.10 레귬 × 우오보 파스타바, 베지테리언 파스타 컬래버레이션 팝업 진행

(Hosted a Vegetarian Pasta Collaboration Pop-up with Uovo Pasta Bar)

2024.05 레귬 × 오브젝티파이, '그라운드 플레이트' 협업 제작

(Produced **The Ground Plate** in collaboration with Objectify)

2024.05 레귬 × 이스트 × 넘은봄, 팜투테이블 컬래버레이션 팝업 진행

(Hosted a Farm-to-Table Collaboration Pop-up with Y'east & Last Spring)

2024.03 레귬 × 밍글스 × 빈호 × 이스트 × 주은 × 마테르 × 레벨제로, 아시아 50 베스트 레스토랑 '뉴 웨이브 오브 서울' 참여

(Participate in **New Wave of Seoul** at Asia's 50 Best Restaurants with Mingles, Vinho, Y'east, Jueun, Mater & Level Zero)

2024.03 레귬 × 제스트 × 주은, 아시아 50 베스트 레스토랑 '서울 나이트' 참여

(Participated in **Seoul Night** at Asia's 50 Best Restaurants with Zest & Jueun)

2023.05 레귬 × 오플레닛, ESG 경영 활성화를 위한 업무협약 체결

(Signed an MOU to enhance ESG management with O.Planet)

2025년 2월 기준
as of February 2025

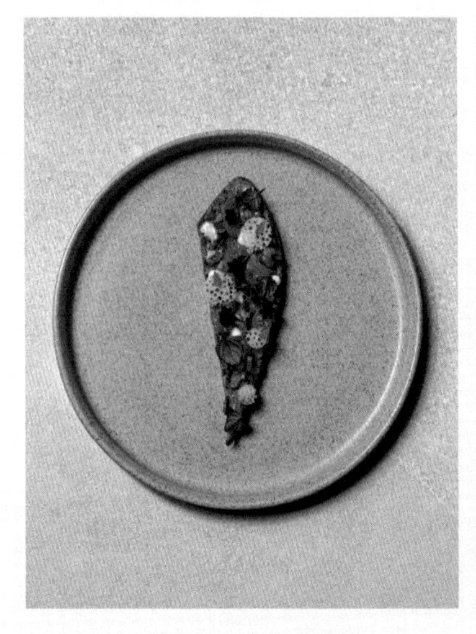

Sung Siwoo

Executive Chef of LÉGUME, a one-Michelin-star vegan restaurant featured in the MICHELIN Guide Seoul & Busan 2025, dedicated to exploring creative plant-based cuisine and sustainable gastronomy.

The VEGAN PANTRY

Your First Step into Everyday Plant-Based Cooking

First edition published	March 20, 2025
Second edition published	April 7, 2025

Author	Sung Siwoo
Translated by	Kim Yoojin
Prep Assistants	Lee Haewook, Koo Mojoon, Lee Wooju
Publisher	Bak Yunseon
Published by	THETABLE Inc.
Plan	Bak Yunseon
Proofreading	Kim Youngran
Design	Kim Bora
Photograph	Shin Dongmin
Sales/Marketing	Kim Namkwon, Cho Yonghoon, Moon Seongbin
Management support	Kim Hyoseon, Lee Jungmin
Address	122, Jomaru-ro 385beon-gil, Bucheon-si, Gyeonggi-do, Republic of Korea
Website	www.icoxpublish.com
Instagram	@thetable_book
E-mail	thetable_book@naver.com
Phone	82-32-674-5685
Registration date	August 4, 2022
Registration number	386-2022-000050
ISBN	979-11-92855-17-2 (13590)

ISBN 979-11-92855-17-2 13590

값 29,000원